Spectroscopic and Microscopy Techniques for Archaeological and Cultural Heritage Research (Second Edition)

Online at: https://doi.org/10.1088/978-0-7503-4846-1

IOP Series in Spectroscopic Methods and Applications

Series Editor
Ashutosh Kumar Shukla, Ewing Christian College, University of Allahabad, Prayagraj, India
Daniel Cozzolino, The University of Queensland, Brisbane, Australia

About the Series

The *IOP Series in Spectroscopic Methods and Applications* aims to develop a library of key texts encompassing the broad range of spectroscopic techniques, instrumentation and fields of applications found in modern science. Some examples of the spectroscopic techniques incorporated include, IR, UV/VIS, Raman, EPR, NMR, fluorescence spectroscopy, force spectroscopy, hyperspectral imaging, neutron scattering, Mössbauer spectroscopy, photoemission spectroscopy and all the various specialist techniques within these broad domains and more. Microscopic characterization techniques may also be included where they are used in conjunction with spectroscopic methods for sample analysis.

The series contains two broad types of approach. Those addressing a particular field of application and reviews the numerous relevant spectroscopic methods applicable to the field, and those that focus on a specific spectroscopic method which will permit a greater in-depth review of the theory and instrumentation technology.

A full list of titles published in this series can be found here: https://iopscience.iop.org/bookListInfo/iop-series-in-spectroscopic-methods-and-applications.

Spectroscopic and Microscopy Techniques for Archaeological and Cultural Heritage Research (Second Edition)

Edited by
Ashutosh Kumar Shukla
Department of Physics, Ewing Christian College, Prayagraj, UP, India

IOP Publishing, Bristol, UK

ISBN 978-0-7503-4846-1 (ebook)
ISBN 978-0-7503-4844-7 (print)
ISBN 978-0-7503-4847-8 (myPrint)
ISBN 978-0-7503-4845-4 (mobi)

DOI 10.1088/978-0-7503-4846-1

Version: 20230501

IOP ebooks

British Library Cataloguing-in-Publication Data: A catalogue record for this book is available from the British Library.

Published by IOP Publishing, wholly owned by The Institute of Physics, London

IOP Publishing, No.2 The Distillery, Glassfields, Avon Street, Bristol, BS2 0GR, UK

US Office: IOP Publishing, Inc., 190 North Independence Mall West, Suite 601, Philadelphia, PA 19106, USA

Contents

Preface

I am happy to present this edited volume entitled *Spectroscopic and Microscopy Techniques for Archaeological and Cultural Heritage Research* to the reader. This is the second edition of the edited volume 'Spectroscopic Techniques for Archaeological and Cultural Heritage Research' published two years ago. This volume not only includes additional spectroscopic techniques but also covers microscopy techniques applied for archaeology and cultural heritage research.

Chapter 1 by Howell G M Edwards introduces the role of analytical instrumentation and techniques and gives the reader the context for the following chapters. Chapter 2 by Meropi Mari and George Filippidis presents nonlinear optical microscopy techniques for cultural heritage studies. Chapter 3 by Eman Osman is an updated version of the same chapter from the earlier edition; it includes case studies and introduces different nondestructive spectroscopic techniques for such applications. Chapter 4 by Wael Badawy, Octavian G Duliu, Mina Magdy, Qunoot Mohammed Hadi, and Sister Serafima presents x-ray fluorescence spectroscopy applications used to investigate works of art. Chapter 5 by Verónica Guilarte presents the principles, applications, and future scope of the potential electron spin resonance (ESR) dating method. This chapter also compares ESR dating with other absolute dating methods and talks about their advantages and drawbacks. In chapter 6, Joseph P Hornak presents ESR spectroscopy applications in the 100–500 MHz frequency range used to study larger objects with cultural heritage significance. Chapter 7 by Shin Toyoda illustrates quantitative assessment of the temperature of past lithic heat treatment through the use of ESR and includes a comparison with thermoluminescence (TL) applications. Chapter 8 by Liviu Tugulan, Mihail Secu, Vasile Bercu, Mihalis Cutrubinis, and Octavian G Duliu presents the determination of the geological age of a particular region based on the potential ESR method and TL. Chapter 9 by Puja Karmakar, Lokanath Mishra, and Monalisa Mishra presents ESR spectroscopy applications used to date mollusc shells, corals, and many other materials.

I sincerely thank John Navas, Senior Commissioning Manager, IOP Publishing for giving me an opportunity to present this updated edition. Additional credit is due to him for connecting me to many experts. I also wish to thank Phoebe Hooper, Editorial Assistant, IOP Publishing for extending all the support received during the different stages of this project. I thank the expert authors for contributing to this volume and making it a unique enriched collection.

Ashutosh Kumar Shukla
Prayagraj, India

Editor biography

Ashutosh Kumar Shukla

Ashutosh Kumar Shukla has more than two decades of physics teaching and research experience. He has had numerous articles and review articles published in peer-reviewed journals. He has also authored textbooks and participated as an editor in more than twenty edited volumes published by reputed publishers, often prepared in collaboration with experts from different countries. He intends to continue pursuing his academic interests, which include the application of spectroscopy in different fields of societal importance.

Contributor biographies

Wael Badawy

Wael Badawy received his PhD in Radioecology and Ecotoxicology from Moscow State University in Moscow, Russia in 2010. He worked at the Egyptian Atomic Energy Authority as an assistant professor. He is now working at the Joint Institute for Nuclear Research as a senior researcher. His main scientific interests are the applications of complex physical techniques in the fields of environmental radiation protection, radiation detection, activation analysis, radiation dosimetry, radioecology, ecotoxicology, archaeology, cultural heritage, and risk assessment. His work has been published in renowned scientific journals. Dr Badawy is a radiation protection expert in non-medical applications and was the Egyptian National Representative to the United Nations Scientific Committee on the Effects of Atomic Radiation (UNSCEAR) until 2018.

Vasile Bercu

Vasile Bercu obtained his BSc and MSc degrees from the University of Bucharest. Between 2002 and 2007, he worked at the Institute for Chemical–Physical Processes of the National Research Council (IPCF-CNR), Pisa, Italy. In 2007, he defended his PhD thesis at the University of Pisa on the study of polymeric and glass former materials using high-field electron paramagnetic resonance (EPR) spectroscopy. In 2007, he became an assistant professor, and since 2019 he has been an associate professor at the Faculty of Physics, University of Bucharest.

After almost six years of high-field EPR spectroscopy experience at the IPCF-CNR (Pisa), Vasile Bercu returned to the University of Bucharest where he set up a new EPR laboratory based on the X band. Using the newly available infrastructure, he has broadened his scientific interests in the use of EPR spectroscopy in food science and cultural heritage.

Mihalis Cotrubinis

Mihalis Cotrubinis graduated from the Faculty of Physics at the University of Bucharest in 1992, and in 2004 he defended his PhD thesis in Physics, also at the University of Bucharest. From his graduation until 2002, he was a scientific researcher at the Institute of Food Chemistry in Bucharest, between 2002 and 2004 he was an invited scientist of the Federal Research Centre for Nutrition and Food, Germany, and since 2004 he has worked as a Senior Researcher at the IRASM Centre for Technological Irradiation, Bucharest.

His main scientific interests are related to the use of ^{60}Co technology for medical, pharmaceutical, and cosmetic products, materials, food, and cultural heritage objects as well as the detection of irradiated foodstuffs by ESR and TL as well as by optically stimulated and photostimulated luminescence.

Octavian G Duliu

Octavian G Duliu received his MS degree in nuclear physics at the University of Bucharest in 1969, a PhD in electron paramagnetic resonance from the V E Lashkaryov Institute of Semiconductor Physics in Kiev in 1973, and a PhD in geosciences from the University of Bucharest in 2012. Since 1969, he has been a member of the Department of Structure of Matter, Earth and Atmospheric Physics and Astrophysics, where he is currently an emeritus professor and PhD adviser. Since 2021, Professor Duliu has worked for the Geological Institute of Romania. Between 1993 and 2019, he held several internships at the Abdus Salam International Center for Theoretical Physics, Trieste, Italy, the University of Patras, Greece, and as leading scientist of the Joint Institute for Nuclear Research in Dubna, Russian Federation. His current research interests are the applications of the physical methods of investigation in physics, geosciences, and biology. Professor Duliu is the author, coauthor, or editor of more than 150 papers published in renowned scientific journals and nine books. Professor Duliu is also winner of the Romanian Academy Gheorghe Munteanu-Murgoci prize as well as that of the Joint Institute for Nuclear Research, Dubna (Russian Federation). He is also a member of the European Geosciences Union.

Howell G M Edwards

Howell G M Edwards is emeritus professor of molecular spectroscopy at the University of Bradford. He received his BA and BSc degrees from Jesus College, University of Oxford, and following research into Raman spectroscopy and its applications, he took his MA and DPhil at Oxford University. After a research fellowship at Jesus College, University of Cambridge, he joined the University of Bradford, where he became the head of Chemical and Forensic Sciences with a personal chair. He has authored and coauthored almost 1400 publications on Raman spectroscopy and its applications, many of which are in the field of cultural heritage artifacts, and he has authored and coauthored 22 books and monographs on ceramics, forensics, and Raman spectroscopy. He is an honorary scientific adviser to the de Brécy Trust for the analysis of oil paintings.

George Filippidis

Dr George Filippidis is a senior staff scientist at the Institute of Electronic Structure and Lasers (IESL), Foundation for Research and Technology (FORTH), Heraklion, Crete, Greece. He received his BSc degree from the Physics Department of the University of Crete in 1995 and his Doctorate degree from the Medical School of the same University in 2000. He performed his postdoctoral studies at IESL-FORTH. His research group activities (in the Non-Linear Microscopy Group) include the *in vivo* elucidation of molecular mechanisms and biological processes through the use of various spectroscopic techniques with an emphasis on advanced microscopic modalities, optical characterization of tissue, nanosurgery, and the implementation of nonlinear image contrast modalities as novel diagnostic tools for cultural heritage studies. He has published 69 articles in international peer-reviewed journals, 31 full-length articles in edited peer-reviewed volumes, 28 conference proceedings and three book chapters. He has been the co-supervisor of three PhDs, eight master's degrees and ten diploma theses. He is the coauthor of one patent. He has given three invited talks in international conferences, a plenary talk, and an invited internet lecture at the Max Born Institute. He has been a member of teaching staff in two master's degree programs.

Verónica Guilarte

Verónica Guilarte graduated at Burgos University (Spain) with an honours award in Chemistry in 2006. She subsequently completed a master's degree in advanced chemistry and her PhD studies, receiving a special doctoral award in science in 2012.

She worked at the National Centre for Human Evolution (CENIEH, Spain) from 2010 to 2016 as the technical supervisor of the Electron Paramagnetic Resonance Laboratory. She is expert in ESR dating and dosimetry using several materials, such as fossil teeth, sedimentary quartz, and corals, among others. In addition, she was the supervisor of the radioactive facility at CENIEH.

From 2014 to 2017, Verónica also worked as a lecturer at Isabel I International University (Spain), and in 2017, she joined Granada University (Spain) as an associate professor in science education.

She has authored 20 publications in international peer-reviewed journals and two book chapters. She has participated in several research and innovative education projects.

She has participated in numerous international and national conferences and research events. In addition, Verónica has been granted visiting professorships and fellowships at The Scripps (San Diego), the University of Cambridge, Institut de Radioprotection et de Sûreté Nucléaire (Paris), and Griffith University (Australia).

Joseph P Hornak

Joseph P Hornak is a professor of chemistry and imaging science at the Rochester Institute of Technology, where he directs the activities of the Magnetic Resonance Laboratory and teaches courses in analytical chemistry and magnetic resonance. His research interests include the development of LFEPR spectroscopy, which is used to study objects with cultural heritage importance. Dr Hornak received his PhD in chemistry from the University of Notre Dame, his MS in physical chemistry from Purdue University, and his BS in chemistry from Utica College of Syracuse University. He is the author of more than 220 scientific works and presentations on magnetic-resonance-related topics.

Puja Karmakar

Puja Karmakar received her BSc degree in zoology from J K College, Sidho Kanho Birsha University, Purulia, West Bengal, India and her MSc degree in zoology from Sidho Kanho Birsha University, Purulia, West Bengal, India. She is currently pursuing a PhD at the Neurodevelopmental Biology Lab, NIT Rourkela, Orissa, India. Her current research interests include the glia and neurons of *Drosophila*. She currently holds the permanent post of assistant professor in zoology at East Calcutta Girls' College, Lake Town, West Bengal, India.

Mina Magdy

Mina Magdy is a senior scientist at the National Museum of Egyptian Civilization (NMEC). He has a bachelor's degree in science, a diploma in biochemistry and physiology, and a diploma in total quality management and quality management systems. He earned his master's degree and a PhD in analytical chemistry from Cairo University. He obtained a lean six sigma black belt certificate from the American University in Cairo. Dr Magdy is interested in the characterization of archaeological materials and the assessment of objects' states. His work has been published in several peer-reviewed journals. He is also engaged in collaborations with different research organizations in the field of cultural heritage. He is a member of many scientific societies.

Meropi Mari

Meropi Mari is a postdoctoral researcher at the Institute of Electronic Structure and Lasers (IESL). She has a BSc degree in Physics (2005) and an MSc degree in Optoelectronics (2008) from the Physics Department of the University of Crete. She received her PhD degree from the University of Edinburgh (2014). Her expertise includes nonlinear microscopy techniques for biological applications but also for art conservation. Her work has been published in scientific articles in international journals and also in conference proceedings. She has recently received funding from the Hellenic Foundation for Research and Innovation in order to set up a two-photon microscope for *in vivo* applications in mice in collaboration with the Institute of Molecular Biology and Biotechnology of the Foundation for Research and Technology Hellas (IMBB-FORTH). At the moment, she is working on an upgrade of this platform to an all-optical system for functional studies by introducing optogenetic manipulations.

Lokanath Mishra

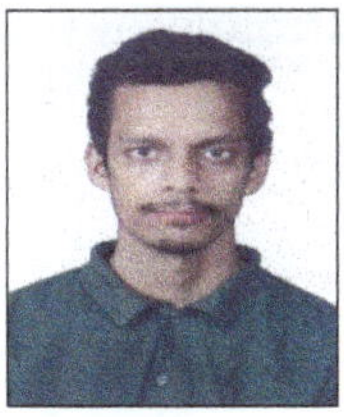

After completing his MSc and MPhil degrees, he gained the CSIR-UGC NET and GATE qualifications. Lokanath Mishra is currently undertaking his PhD at the Neurodevelopmental Biology Lab, Department of Life Science, National Institute of Technology, Rourkela, Odisha, India.

Monalisa Mishra

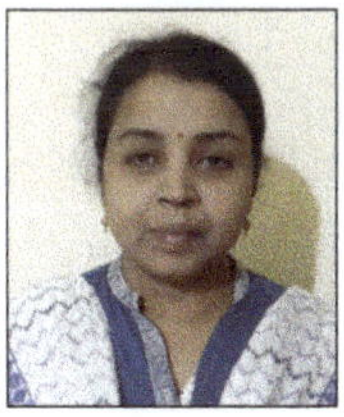

Monalisa Mishra was awarded her MSc by Khallikote (Autonomous) College, achieving Zoology honors in 1999. She completed her MPhil at Berhampur University in 2001 with a specialization in cytogenetics. In April 2004, she joined Jacob's University, Bremen, Germany as a visiting scholar. In September 2004 she joined the same institution as a PhD student under the supervision of Professor Dr V B Meyer-Rochow. The subject of her PhD was 'Eye ultrastructure under various photic condition.' She finished her PhD in June 2007 and in July moved to the School of Medicine, University of San Diego, California, for her first postdoctoral position. There, she worked on Usher's syndrome (an eye disease) in patients and also using a mouse model. She completed her second postdoctoral assignment (April 2008) at Indiana University, Bloomington. There, she worked on *Drosophila* (the fruit fly) eyes. In April 2010, she joined MPI-CBG, Dresden, Germany for her third postdoctoral position. In July 2012, she joined BITS, Pilani as an assistant professor. In May 2014, she moved to NIT Rourkela and has worked since then using *Drosophila* as a model animal. Currently, her lab is working on how genes affect the eye and mechanosensory organ development.

Qunoot Mohammed Hadi

Qunoot Mohammed, from Babylon, Iraq, received a bachelor's degree in physics in 2010 and a master's degree in physics in 2016, both from the College of Education for Pure Sciences, University of Babylon, Iraq. In the same year, she received a patent from the Ministry of Planning in Iraq concerning the fabrication of a solar cell made from titanium dioxide doped with magnesium oxide and deposited on a silicon substrate by the pulsed laser deposition method. She has been involved in the applications of nanomaterials in the manufacture of solar cells. In 2018, she obtained the scientific title of assistant lecturer after passing the Teaching Methods course and the teaching validity test at the Department of Physics, College of Education for Pure Sciences, University of Babylon. Since 2019, she has been a PhD candidate at the Doctoral School of Physics of the University of Bucharest, where she is preparing her PhD thesis concerning the investigation and analysis of archaeological materials.

Eman Osman

Professor Dr Eman Osman is a professor of textile metrology at the Chemical Metrology Division, National Institute of Standards, Egypt. She has published a lot of research work, book chapters, and encyclopedia entries in the fields of:

- natural dyeing and the use of non-destructive techniques in cultural heritage (textiles and manuscripts)
- nanotechnology and its uses in textile and archaeological applications
- the electrospinning technique and its application in medical textiles
- the chemical, physical, thermal, and mechanical testing of textiles

In addition, Professor Osman has supervised and reviewed many MScs and PhDs in the fields of archaeology and textile modification. Dr Osman currently occupies the position of head of the Chemical Metrology Division at the National Institute of Standards.

Mihail Secu

Mihail Secu received his physics diploma from the University of Bucharest in 1992 in the field of optical spectroscopy. In 1992, he started working at the National Institute of Materials Physics, where he defended his PhD thesis on the investigation of radiation defects in various optical materials in 2000. He then completed a three-year research stay abroad as a researcher at University of Paderborn, Germany in the field of x-ray storage phosphors, followed by five months as an invited research scientist at the

University of Groningen, the Netherlands in 2005, working in the field of the optical dating of sediments. In the following years, he further expanded his research into the synthesis methods and optical characterization of nanostructured materials and radiation-induced defects.

Sister Serafima

Since 1985, Sister Serafima has been a nun at the One Wood Monastery, Valcea County, Romania. In 2009, she graduated at the Faculty of Orthodox Theology of the University of Pitesti; in 2011, she received an MS degree in the history and theory of art at the department of Conservation and Restoration of Works of Art at the National University of Arts in Bucharest; and in 2017, she completed a PhD in physics in the Applications of Physical Techniques of Investigation in the Expertise of Ancient Romanian Icons. Since 2007, she has worked as a tempera painting restorer; in 2011 she was certified as an expert restorer in the tempera technique on wooden supports and polychrome wood; while since 2021, she has been the director of the Restoration Research and Conservation Center–Saint Hierarch Calinic of Cernica in the Archdiocese of Râmnic. In 2018 she was appointed as an expert in physicochemical investigations. Between March 2019 and September 2021, Sister Serafima acted as the mother superior of Sărăcineşti Monastery, County Vâlcea, Romania. Sister Serafima has coauthored several scientific papers published in well-known journals. She is also the member of the Conservators and Restorers Association of Romania (http://www.acrro.ro).

Shin Toyoda

Shin Toyoda is a professor at Okayama University of Science. His current research interests include electron spin resonance dating dosimetry and sediment provenance. Dr Toyoda received his BSc and MSc degrees from the Geophysical Institute, University of Tokyo, Japan. He received his PhD from Osaka University, Japan. Dr Toyoda has several publications to his credit. Dr Toyoda has been a recipient of an award for the encouragement of young scientists given by the Geochemical Society of Japan.

Liviu C Tugulan

Liviu C Tugulan graduated at the Physics Faculty of the University of Bucharest; in 2015, he was awarded a PhD in Physics by the University of Bucharest. Since 2011, he has worked as a scientific researcher at the Horia Hulubei National Institute for R&D in Physics and Nuclear Engineering (IFIN-HH) in the field of ionizing radiation metrology and radioactive waste management as well as the monitoring of the controlled areas of the Radioactive Waste Treatment Plant (STDR) and the National Radioactive Waste Repository (DNDR).

He is a specialist in the high-accuracy determination of the level of activity of environmental radionuclides (including tritium) in water by the liquid scintillation method; the characterization of solid and liquid waste using gamma-ray spectrometry, scintillation analysis, and global alpha/beta measurements; and the determination of the contents of major rock-forming elements in cement, geological samples, soils, etc. by x-ray fluorescence.

List of contributors

Wael Badawy
Joint Institute for Nuclear Research, Russian Federation and Egyptian Atomic Energy Authority, Russia

Vasile Bercu
University of Bucharest, Bucharest, Romania

Mihalis Cutrubinis
Horia Hulubei National Institute for R&D in Physics and Nuclear Engineering, Romania

Octavian G Duliu
University of Bucharest, Bucharest, Romania
and
Geological Institute of Romania, Bucharest, Romania

Howell Edwards
University of Bradford, Bradford, UK

George Filippidis
Institute of Electronic Structure and Laser, Heraklion, Greece

Verónica Guilarte
Universidad de Granada, Campus de Melilla, Spain

Joseph Hornak
Rochester Institute of Technology, Rochester, NY, USA

Puja Karmakar
NIT Rourkela, Odisha, India

Mina Magdy
National Museum of Egyptian Civilization, Cairo, Egypt

Meropi Mari
Institute of Electronic Structure and Laser, Heraklion, Greece

Lokanath Mishra
NIT Rourkela, Odisha, India

Monalisa Mishra
NIT Rourkela, Odisha, India

Qunoot Mohammed Hadi
University of Babylon, Babylon, Iraq

Eman Osman
National Institute for Standards, Giza, Egypt

Mihail Secu
National Institute of Materials Physics, Romania

Sister Serafima
One Wood Monastery, Frânceşti Village, Romania

Shin Toyoda
University of Science, Okayama, Japan

Liviu Tugulan
Horia Hulubei National Institute for R&D in Physics and Nuclear Engineering 30, Romania

Chapter 1

Introduction

Howell G M Edwards

This chapter first comprises a general survey of the role of analytical instrumentation and techniques in the cultural heritage preservation of artifacts from archaeological excavations and museum collections. The selection of an appropriate technique based upon elemental and molecular spectral signatures is crucial to achieving an understanding of the chemical and mineralogical composition of an artifact, the degradation that has occurred during its burial or storage and whether this is still ongoing, and the technologies of production and the sourcing of raw materials used in its construction and synthesis. The determination of the age of an artifact and the detection of fakes and forgeries are also integral parts of the analytical interrogation of a cultural heritage artifact. A brief survey of the following chapters and techniques comprising this volume is also undertaken.

1.1 Setting the scene

The input of scientific research into the elucidation of problems in archeology, museum artifact curation, and cultural heritage materials at the arts/science interface is receiving greatly increased attention from analytical scientists, and the dissemination of knowledge is reaching a wider audience that is appreciative of the novel information that can be accessed through such interdisciplinary project applications. From simple beginnings at the fringe of analytical science several decades ago, more complex problems and questions are now being addressed using a battery of modern analytical equipment and instrumentation, often used in a combinatorial sense to elicit elemental atomic and molecular spectral data from a wide range of materials which comprise artworks and functional artifacts from our human history which feed into a holistic assessment and a better understanding of the oft-forgotten processes and procedures used by our ancestors. In this respect, the vital assimilation of scientific data, historical documentation, provenancing, and archaeological

doi:10.1088/978-0-7503-4846-1ch1

expertise and connoisseurship must be considered holistically to arrive at an objectively evaluated assessment of the cultural heritage of a specimen or artifact from an archaeological excavation, gallery, or a museum collection.

1.2 Analytical science: the exposure of fakes and forgeries

In a forensic archaeological sense, the key questions that usually require an answer or some contribution from an analyst regarding a cultural heritage specimen are convolved in some way with the following dictats: what is it composed of (what is in it, elementally, biochemically, and mineralogically), how much of each component is there in the artifact, where did these originate and perhaps the most difficult question of all, why is the detected material there? The first three of these requirements are addressed quite uniformly by analytical data from spectroscopic and microscopic analyses that have preferably been accomplished using the combination of several techniques to generate the specific information required. However, the fourth question is rather more difficult and its answers often depend upon associated material found in an excavation and the effects of environmental degradation; in particular, the bacterial biodeterioration of biologically derived materials and artifacts is significant in this context. It must be appreciated that the earliest applications of analytical science to archaeological or museum artifacts were necessarily destructive of the samples concerned: in several instances, a significant part of the specimen (the analyte) had to be sacrificed to provide the scientific analytical information required. The earliest analysis of Chinese porcelains made to ascertain their mineralogical composition was the work of René-Antoine Ferchault de Réaumur [1–3] in Paris in the 1730s, followed by the first analytical study of archaeological material by Sir Humphry Davy [4] in London in 1815, who analyzed the pigments on shards of the newly excavated Roman wall paintings preserved at Pompeii. Davy specifically deplored the fact that he had to destroy the specimens to accomplish his analyses. An opposite example of this was the groundbreaking work undertaken by Herbert Eccles and Bernard Rackham [5] to determine the compositions of early English porcelains in the Lady Charlotte Guest-Schreiber Collection of the Victoria and Albert Museum in South Kensington, London, where the sacrifice of a whole piece of porcelain, which was admittedly normally already damaged anyway, was deemed necessary to accomplish the task and provide the chemical information required. In his early analytical studies of porcelains, Sir Arthur Church [6] also sacrificed the only known surviving piece of Brancas-Lauranguais porcelain from 1760, made in London by a French aristocratic émigré, that had survived the conflagration and inferno which destroyed the Alexandra Palace exhibition in London in June 1873 [7] in his determination of its elemental composition. These compositional determinations involved the wet chemical digestion of significant portions of often unique specimens in totally destructive procedures that would most certainly not be countenanced today by museum curators or collectors. This has steadfastly given rise to the now-favored adoption of noninvasive analytical sampling techniques which do not require the physical or chemical removal of material from the particular artifact concerned. An intermediate situation now applies for several techniques in which only micro samples are permitted to be removed from the specimen, in the range of microgram or picogram (10^{-6} g and 10^{-9} g, respectively) quantities, from a visibly inobvious or hidden part of the artifact—this is then referred to as a 'minimal sampling' procedure [8, 9]. In some cases, the interrogation of an artifact using noninvasive techniques and portable on-site instrumentation is used as a screening operation which is followed by selective minimal sampling for further laboratory analysis of special areas of interest.

Some of the earliest analytical studies of cultural heritage artifacts were directed at ascertaining the composition and identification of the decorative pigments in illuminated historiated vellum medieval manuscripts, and here a distinct advantage was realized by the provision of minute specimens that had become detached from their original historiated placements in the text and had then been accumulated in the bindings of the leaves of the manuscript concerned. This provided a rich source of microspecimens for subsequent microscopic and spectroscopic analysis. In this context, the concept of the 'authentication' of a cultural heritage artifact arose, since the identification of an chronologically out-of-context pigment would most certainly have predestined the assessment of the artifact to the category of being a fake or a forgery. Such is the case of the *El Faussaire Espagnol*, the Spanish Forger, who operated in The Netherlands and Belgium in the late 19th and early 20th century and who faked several hundred purportedly medieval miniatures on Biblical themes, many of which (it was then estimated by Wilhelm Voelkle that there were about 150 of these forgeries extant; this has now been revised to a number in excess of 350 forgeries!) now reside in museum collections worldwide. An illuminated historiated D illustrating the 'Visitation' on a medieval choir book, which has been shown to be a fake attributed to the Spanish Forger, is shown in figure 1.1. The work of the Spanish Forger had nevertheless been deemed genuine by connoisseurs historically, and although some suspicions regarding several artifacts had been raised on stylistic grounds, these were eventually exposed through the detection of out-of-context pigments used in their construction that did not exist in the 14th and 15th centuries, such as chrome yellow (lead (II) chromate, $PbCrO_4$), Prussian blue (potassium hexacyanoferrate, $K_4Fe(CN)_6$), and anatase (titanium (IV) oxide, TiO_2). In 2009, Burgio *et al* [10] demonstrated through Raman spectroscopic and x-ray fluorescence analysis that five 'medieval' miniatures in the collection of the Victoria and Albert Museum, South Kensington, London, which were attributed to the Spanish Forger, contained synthetic pigments which did not exist in the 14th and 15th centuries, such as ultramarine blue (first synthesized in 1828) and Scheele's green (first synthesized in 1775). Coomans and de Maeyer [11] have produced an extensive *catalogue raisonné* of the works of the Spanish Forger, building upon the earlier study of Voelkle and Wiecke [12].

A codicil to the provenancing of the artifacts of cultural heritage by analytical science has been expressed by Professor Virginia Orna [13] as follows:

> *'Science cannot authenticate an artwork but can only expose a fake.'*

Figure 1.1. An historiated 'D' from a choir book depicting the 'Visitation,' purportedly medieval, but now attributed to the Spanish Forger, dating from the late 19th century to the 1920s. Tempera and gold on parchment. Courtesy of the Metropolitan Museum of New York, Gwynne M. Andrews' bequest, 1930. Public domain, CC0.

This is still adopted as a mantra and held to apply generically, but it has been shown more recently that scientific analysis has demonstrated that some very unusual porcelain artifacts that have hitherto been consigned to the category of being fakes or of questionable origin stylistically because they were not considered to be the normal products of manufacture have now had their attribution confirmed as genuine exemplars [8, 9]. Hence, the authentication of an artwork is still possible using analytical science, even when stylistic expert connoisseurship dictates otherwise.

Paul Craddock [14] has assimilated a wide range of examples of the applications of scientific analysis to cultural heritage artifacts involving metals, ceramics, glass, enamels, stone, sculpture, paintings, prints, documents, gemstones, jade, and organic materials whilst working as a scientist in the British Museum Research Laboratory, London. Specifically addressing the topic of fakes and forgeries of artifacts, Craddock describes the key analytical techniques and instrumentation that have been deployed to study specimens and considers the effect of recorded or unrecorded restoration and other modifications which may have been made in antiquity which can create problems in the interpretation of analytical data obtained from present-day studies.

1.3 The effect of degradation upon the analytical interpretation of data obtained from cultural heritage artifacts

The effect of an analytical experiment undertaken with a cultural heritage artifact that has been subjected to the deterioration of its matrix substrate or an applied decoration upon the interpretation of data can give rise to pitfalls if it is not recognized as such. Such a deterioration may have occurred through the incompatible use of neighboring pigments or binders, such as the interaction between an egg tempera and a lead white ($Pb(OH)_2{\cdot}PbCO_3$) pigment on a vellum manuscript, which generates an unsightly black discolouration in the painted areas from the formation of lead sulfide (PbS, galena) by a reaction between the sulfur in the egg tempera and the lead (II) ions of the white pigment. Similarly, the absorption of radiation in the visible region of the electromagnetic spectrum at around 520 nm causes a transition from realgar (As_2S_3) to pararealgar (As_4S_4) pigments and a resulting change in color from yellow to orange [15]. Similarly, the absorption of radiation across a wider wavelength range in the visible spectrum constitutes an irreversible change in color of cinnabar (mercury sulfide, HgS) from red to black arising from the conversion of the alpha to the beta polymorph of the mineral pigment; this has been observed when attempts have been made to clean old oil paintings using laser radiation. Thermal energy converts the bright blue azurite ($CuCO_3$) and green malachite ($Cu(OH)_2{\cdot}CuCO_3$) pigments into brown tenorite (copper (I) oxide, Cu_2O); this effect has been observed during the restoration of the Renaissance frescoes in the church of Sant Joan del Mercat in Valencia, which were destroyed by fire in the Spanish Civil War [16].

Aside from these inorganic mineral species, the biodeterioration and environmental degradation that can be sustained by biomaterials of cultural heritage interest are also major concerns: the degradation of textiles, human and animal skeletal remains, resins, waxes, and plant materials is an ongoing problem for museum conservators who need analytical data to determine the course of action required for the preservation of a decaying artifact and, more importantly for

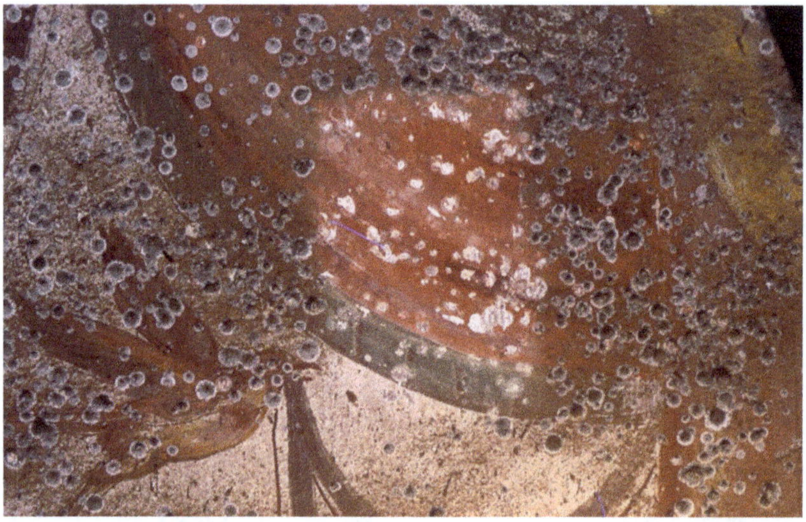

Figure 1.2. A section of the biodeteriorated Renaissance frescoes in an inner courtyard of the Palazzo Farnese, Caprarola, Rome, showing extensive lichen invasion by *Dirina massiliensis* forma *sorediata*; areas of spallated white de-pigmented surface are evident where the lichen thalli have become detached from the substrate. Lichen hyphae have penetrated into the substrate marble.

bacterial biodeterioration, to ascertain whether the deterioration is still ongoing or not and whether it requires the input of immediate attention to arrest and prevent further decay whilst the artifact is in museum storage. This is especially noticeable for the preservation of human skeletal remains and mummified tissues as well as for textiles and resins. An excellent example of the effect of environmental biodeterioration is provided by the analytical studies of the lichen invasion of *Dirina massiliensis* forma *sorediata* of the 16th century frescoes in the courtyard of the Palazzo Farnese, Caprarola, Italy, shown in figure 1.2 [17, 18]. Here, the coverage of the surface is very extensive and in some places it accounts for 80% of the visible artwork; the lichen has adopted a strategy whereby the heavy metals used in the pigments it is colonizing are immobilized and the production of its waste oxalic acid can be seen to appear as calcium oxalate mono- and dihydrates, whewellite and weddellite, $CaC_2O_4{\cdot}H_2O$ and $CaC_2O_4{\cdot}2H_2O$, respectively, which can be discerned as white encrustations at the edges of the gray lichen thalli. The search for calcium ions with which to complex the waste oxalic acid from the lichen metabolism has resulted in the lichen hyphae penetrating through the decorative surface and attacking the marble substrate. Conservators have estimated that one kg of $CaCO_3$ per square meter of substrate has been so converted into calcium oxalate, and chemical spectroscopic analysis has detected the presence of the lichen hyphae extending some 15 mm into the underlying marble substrate.

1.4 Analytical techniques used for cultural heritage artifacts

The range of analytical techniques that have been adopted to study artifacts of relevance to our cultural heritage preservation is now quite formidable, and many research papers cite the combination of more than one technique to interrogate specimens. Generally, one can divide the techniques into three categories of information provision according to their primary purpose, namely elemental analysis, molecular analysis, and dating. Chemical compositional data can be derived from the first two of these categories, which include mineralogical information particularly relevant for the assessment of raw materials that have been used in the creation of the artifacts themselves. The techniques can therefore be broadly subdivided as follows:

Elemental and molecular: x-ray diffraction (XRD), scanning electron microscopy (SEM)/energy dispersive x-ray spectrometry (EDAXS), x-ray fluorescence (XRF), Raman spectroscopy, surface-enhanced Raman spectroscopy (SERS), infared (IR), ultraviolet spectroscopy (UV), gas chromatography–mass spectrometry (GC–MS), electron paramagnetic resonance (EPR)/electron spin resonance (ESR), nuclear magnetic resonance (NMR);
Dating: ESR, thermoluminescence (TL), accelerator mass spectrometry (AMS) (C^{14});
Mineralogical: XRD, Raman, IR;
Trace element detection: XRF, SEM/EDAXS, EPR/ESR;
Microscopy: optical microscopy (OM).

A list of the analytical techniques that have been applied to cultural heritage materials is given in table 1.1, in which 35 techniques are tabulated: all of these are cited in the literature as having been used to characterize artifacts in museum collections and from archaeological excavations. It is important to recognize that each of these techniques essentially provides discrete information that is often complementary and that not all techniques are suitable for every determination in particular: for example, radiocarbon dating (AMS) is not an appropriate technique for adoption in the case of artifacts that are prehistoric and older than about 50 000 years, since this represents approximately ten half-lives of the radioisotope carbon 14 (C^{14}), at which point the isotope percentage has been reduced to about a fraction of 0.001 of its original quantity, after which it becomes an unreliable analytical measurement. Needless to say, for this determination the presence of a carbonaceous material that gives CO_2 upon burning is essential, so it is classified as a destructive technique because the consumption of the carbonaceous specimen is irreversible. Likewise, the use of infrared and Raman spectroscopy to detect the presence of gold in a pigment on manuscripts is inappropriate, as these are molecular techniques and gold is an atomic metal that does not possess Au–Au bonds that can be detected molecularly; similarly, the purple of Cassius used as a pigment in Renaissance decoration comprises colloidal atomic gold and is not detectable using these techniques. Hence, it can be stated that the appropriate selection of a technique that is able to obtain useful data from a specific artifact is vitally important: for example, the chemical composition of an organic material can be determined satisfactorily using GC–MS, IR, Raman, SERS, UV, and NMR spectroscopies, but less well with elemental techniques such as SEM/EDAXS, EPR, and XRF

Table 1.1. Physical techniques used for the analysis of cultural heritage artifacts.

Acronym	Brief description	Purpose/type of artifact studied
AMS	Accelerator mass spectrometry	Radiocarbon dating (C^{14}); organic materials—textiles, tissue, wood
ATR	Attenuated total reflectance spectroscopy	Infrared technique for solid surfaces
CL	Cathodoluminescence	Mineralogy, geology; glass, ceramics
EDX	Energy-dispersive x-ray spectrometry	Quantitative analysis of solids; ceramics, pigments, glass, textiles
EMPA	Electron microprobe analysis	Elemental analysis of solids
EMR	Emission radiography	Examination of the interior of an artifact
ES	Emission spectrography	Elemental analysis
ESR	Electron spin resonance	Unpaired paramagnetic electrons, transition metals; dating
EPR	Electron paramagnetic resonance	Alternative name for ESR
FTIR	Fourier transform infrared spectroscopy	Molecular signatures; solids, liquids, gases
FTRS	Fourier transform Raman spectroscopy	Molecular signatures; solids, liquids, gases
GC–MS	Gas chromatography–mass spectrometry	Gases, volatile organics
HPLC	High-performance liquid chromatography	Organics and metallo-organics in solution
ICP-AES	Inductively coupled plasma atomic emission spectroscopy	Elemental analysis
ICP-MS	Inductively coupled plasma mass spectrometry	Elemental analysis of organics, molecular structures
LAICP-MS	Laser ablation–ICP-MS	Elemental analysis of volatilized materials
MagSus	Magnetic sucsceptibility	Presence of transition metals in unusual oxidation states
MS	Mass spectrometry	Structures and presence of volatile organics
NAA	Neutron activation analysis	Elemental analysis; solids
NMR	Nuclear magnetic resonance	Molecular structures of complex materials
OM	Optical microscopy	Mineralogy, petrology, sections, pigments
PIXE	Proton-induced x-ray emission	Elemental analysis
RBS	Rutherford backscatter spectroscopy	Elemental analysis
RM	Raman microscopy	Molecular analysis using a microscope
RRS	Resonance Raman spectroscopy	Selective enhancement of molecular species present
RS	Raman spectroscopy	Molecular analysis, mineralogy
SEM	Scanning electron microscopy	Surface elemental analysis
SERS	Surface-enhanced Raman spectroscopy	Molecular enhancement using silver solutions

SERRS	Surface-enhanced resonance Raman spectroscopy	Molecular enhancement using silver solutions and laser excitation
SIA	Stable isotope analysis	Dating; analysis of archaeological tissues
TEM	Transmission electron microscopy	Elemental analysis
TL	Thermoluminescence	Dating pottery and ceramics, determining production temperatures achieved
XRD	X-ray diffraction	Crystal structures, mineralogy
XRF	X-ray fluorescence	Elemental analysis, minor components
UV	Ultraviolet spectroscopy	Molecules absorbent in the ultraviolet region; structures

unless contamination with transition metals is evident. Although the analysis of coinage metals is not normally undertaken using molecular spectroscopic techniques, the detection of their corrosion products, which are molecular or molecule–ionic entities, is possible; examples of such corrosion products include basic copper acetate (verdigris) and silver salts.

By far the most useful and well-used combination of compositional techniques involves elemental and molecular determinations, such as Raman/IR/SERS and SEM/EDAXS/XRF. Optical microscopic techniques are essential providers of information about the mineralogical and petrological phases of matter, especially for thin sections of a heterogeneous specimen. They are also used for the detection of pigments in admixture and an assessment of their fineness of grinding and production, such as the effectiveness of mixing of raw materials used in ceramics production and the determination of mineralogical phases in porcelains, stonewares, and pottery formed at elevated temperatures in kilns. The use of high-powered optical microscopes to examine the crystal structures of powdered pigments has revealed that the anatase (titanium (IV) oxide) found in the black ink on the Vinland Map (figure 1.3), which has been radiocarbon dated to 1432 ± 11 i.e. within one standard deviation of the observed bull's head watermark of 1434 on the associated Tartar Relation, comprises rounded spherical particles that are more typical of modern synthetic analogs, although there is some ongoing controversy about this interpretation when comparisons are made using the PIXE and Raman techniques [19, 20]. The importance of the Vinland Map in cultural heritage studies relates to its existence, if genuine, some 60 years prior to the voyage of Christopher Columbus and his discovery of Hispaniola in 1492; several analytical techniques have been adopted for the Vinland Map, but their interpretation is somewhat still controversial. There are proponents and opponents of its acceptance as genuine, which are centered upon the composition of the iron gall ink used in the construction of the coastline of north America, christened Vinland by the Norse explorers. It forms an analytical case study in its own right and the interested reader is referred to chapter 7.2 in reference [20] for a more detailed survey [21].

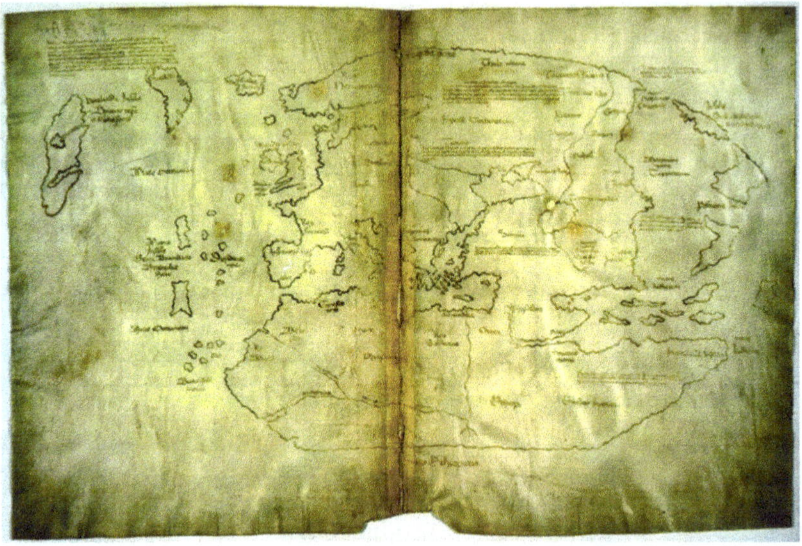

Figure 1.3. The Vinland Map. Courtesy of the Beinecke Library, Yale University, Connecticut, USA. Contrast and saturation digitally enhanced by J H McCulloch: http://www.econ.ohio-state.edu/jhm/arch/vinland/vinland.htm.

It is beyond the scope of this text to itemize the use that has been made of each technique for the characterization of cultural heritage materials and artifacts, and the interested reader is therefore referred to the more extensive bibliography provided in table 1.2 where case studies and exemplars better illustrate the application of analytical techniques that have been adopted either singly or in combination for the characterization of artifacts and materials of relevance to cultural heritage in museum collections or from archaeological excavations. A perusal of this literature will demonstrate in some detail the successful analytical applications, the types of artifact or cultural heritage material that have been studied, and the novel information that has been derived from the application of a range of appropriate techniques. For ease of further reference, full details of the texts cited in table 1.2 are given in the list of references [8, 9, 14, 20, 22–29] here and a survey article digest of the type of molecular spectroscopic analytical data forthcoming from the Raman spectroscopy (RS) analytical technique for a variety of exemplars is given in [30].

1.5 The degradation of cultural heritage artifacts

It is clear that analytical science can make a major contribution to the preservation of cultural heritage artifacts in the assessment of their degradation and in the identification of issues surrounding their ongoing storage or their post-archaeological excavation treatment. There are now many papers cited in the literature that deal with this aspect, particularly regarding the deterioration of textiles, human skeletal remains, pigments on manuscripts, wall paintings and frescoes, and the

Table 1.2. Books describing the application of analytical techniques to cultural heritage artifacts.

Authors	Year	Title	Publisher	Major techniques
Craddock	2009	*Scientific Investigation of Copies, Fakes and Forgeries*	Butterworth-Heinemann/ Oxford	Several
Sgamellotti, Brunetti, Miliani	2014	*Science and Art: The Painted Surface*	RSC/ Cambridge	Several
Edwards, Chalmers	2005	*Raman Spectroscopy in Art and Archeology*	RSC/ Cambridge	RS
Edwards, Lewis	2001	*Handbook of Raman Spectroscopy*	Marcel Dekker/ New York	RS
Edwards	2020	*18th and 19th Century Porcelain Analysis*	Springer/ Dordrecht	RS, SEM, EDAXS, XRF
Edwards	2022	*Porcelain Analysis*	Springer/ Dordrecht	RS, SEM, EDAXS, XRF
Edwards, Vandenabeele	2023	*Raman Spectroscopy in Cultural Heritage*	Springer/ Dordrecht	RS, IR, SEM, EDAXS, XRF, and Colombian AMS
Ciliberto, Spoto	2000	*Modern Analytical Methods in Art and Archeology*	Wiley/ Chichester	Several
Pollard, Heron	1996	*Archaeological Chemistry*	RSC/ Cambridge	Several
Chalmers, Edwards, Hargreaves	2012	*IR and RS in Forensic Science*	Wiley/ Chichester	RS, IR, and others
Edwards	1999	*Encyclopedia of Spectroscopy and Spectrometry*	Academic/ London	RS, IR, and others
Vandenabeele, Edwards	2019	*RS in Archeology and Art History*	RSC/ Cambridge	RS and others

identification of regions on artifacts where previous improper restorative procedures, perhaps involving incorrect adhesives, have been undertaken that have now potentially compromised the integrity of the artifact itself.

1.5.1 Specific themes related to the degradation of cultural heritage artifacts

1.5.1.1 Hierarchical pigments

A major contribution of analytical science to pigment characterization is the identification of pigment mixtures that were compounded by ancient scribes or artists to produce a range of colors from a necessarily limited selection of available minerals. In particular, the admixture of red and yellow pigments to make orange, blue and yellow pigments to make green, red and blue pigments to make purple, and other mixtures to effect color tonal changes such as red and black to make a darker color and red and white to make a lighter color (pink) [24, 30]. More subtle usage of mineral pigments reflected their economic circumstances or a hierarchy of depiction: for example, at the Church of SS Cosmo y Damien at Basconcillos del Tozo on the Camino de Santiago in Spain, a 13th century depiction of Christ in Majesty flanked by the Virgin Mary and an angel on a wall painting is supported by St Peter and a vision of Hell with demons and monsters devouring people [31]. The red robe of Christ is pure cinnabar (mercury (II) sulfide), a very expensive pigment, whereas that of St Peter is found to be cinnabar heavily adulterated with the cheaper hematite or iron ochre, Fe_2O_3. The blue cloak of the Virgin Mary is composed of finely ground lapis lazuli, an expensive pigment which was obtained from mines in what is now Afghanistan. Often, the signatures of minor components have been used to locate the origin of these minerals, such as cinnabar from Almaden or a more local mine at Tarna in northern Spain; more recently, such signatures have been used to identify the source of lapis lazuli in a 20th century artifact as being a Chilean mine rather than the ancient mines in Afghanistan; this arose from alternative sourcing necessitated by an interruption of supplies.

1.5.1.2 Biological materials

Perhaps the most difficult artifacts to study analytically are those comprising biological materials or biomaterials which have been obtained from plants and animals: these can be subdivided into biological artifacts of cultural heritage importance; human and animal skeletal remains, including tissues, such as mummified bodies (which may have been ice-mummified or deliberately treated, such as the Egyptian dynastic mummies), hair, nails, bone, ivories, teeth, feathers, and fur; textiles; preserved organs; natural extracts of plants used as unguents; and aromatic resins such as frankincense, myrrh, ambers, copals, gums, and waxes. Understanding the degradation that such materials have experienced, particularly the fossilized resins such as ambers and copals, is a vital component of their objective assessment for their conservation and restoration and analytical science has a major role to play in this process, as will be seen below. We have already briefly considered the effects of biological colonization and invasion of geological or biomaterial substrates, and special analytical techniques are now available that can be used to assess these qualitatively and quantitatively, including bioanalytical techniques. The spectroscopic techniques already mentioned can also be used to determine the presence of key spectral signatures such as those of chlorophyll, carotenoids, scytonemin, and phycocyanins which are produced during the survival strategies of bacteria that are colonizing biomaterials, thereby assisting in their identification and assignment for their effective neutralization and removal for the preservation of the artifact concerned. A particularly appropriate example of the input of analytical data to a degraded resin is given by the study of a red colored molded resin *tembeta* [32] which was used as a lip plug by Amazonian tribes in Brazil; the question as to whether the damaged *tembeta* found archaeologically was originally coated with a red pigment or if this had occurred during degradation of the resin was answered by analytical RS microbe spectrometry and was fundamental information for the conservators charged with the restoration of the lip plug.

Other resins that have been studied analytically are ambers and copals, fossilized pine resins of varying maturity which are often several million years old that occur geologically in many parts of the word. As in the case of pigments, where elemental and mineralogical analytical data have contributed to sourcing data which have been useful in establishing ancient trade routes, the specific molecular signatures of fossilized ambers and copals have been indicative of their origins, such as Mexican copal, Polish pit amber, and Baltic and Sicilian ambers. Much fundamental research has been undertaken into the characterization of the definitive analytical molecular signatures of ancient resins arising from their particular depositional environments, which then provide a database for the assignment of their geographical origins and in some cases give an estimate of their degree of maturity [33, 34].

An example of the analytical input to cultural heritage artifacts of human skeletal remains, namely the case of the Alpine Iceman can be considered here. In September 1991, two hikers were making their way across the Similaun Pass in the Austro-Italian Alps near the border between Austria and Italy when in the fading light they came across a body encased in a melting glacier. This transpired to be Ötzi, the Alpine Iceman; radiocarbon dating of his skin tissue showed him to be about 5200 years old (AMS C^{14} measurements were in the range between 5320 and 5020 years ago, i.e. 3320–3020 BCE) [35]. Only one small skin tissue sample was taken for the radiocarbon experiment and it was necessary that other analytical work did not compromise this: noninvasive FTRS was used to characterize the skin tissue prior to its destruction in the AMS C^{14} determination; this showed that degradation to adipocere had not set in and that the skin was remarkably well preserved with little evidence of deterioration, with the proteinaceous and lipid components being particularly well defined [36, 37], although spectral deconvolution procedures needed to be applied to resolve the complex features of the human skin.

Bodies found in archaeological excavations are usually clothed, and the textiles can frequently provide important information about technological achievements of the time and the status of the individual at burial. Many examples in the literature illustrate the degradation suffered by textiles, which may additionally have been subjected to treatment using resins, waxes, and unguents at burial. An example of the degradation of the linen textile fibers in a specimen from a burial in the Middle East at around 600 CE has been investigated [38] and the breakdown of the cellulosic C–O–C linkages in the glycosidic polymeric structure of the linen has been demonstrated by RS.

Another example is provided by the degradation suffered by the reinforced linen sailcloth of the foretopsail of *HMS Victory*, Admiral Nelson's flagship at the Battle of Trafalgar in October 1805 when the Royal Navy broke the line of the combined Franco-Spanish fleets under Admirals Gravina and Villeneuve off Cape Trafalgar. FTRS textile analysis of this sail, the only sail to survive the battle on both sides, was carried out and the extent of the degeneration of the fibers revealed, which assisted in the conservation of the *Victory* sail and its subsequent display for the 200th anniversary of the battle at Portsmouth Historic Naval Dockyard in October 2005 [39]. A painting by J M W Turner of the *Victory*, now in the National Maritime Museum, Greenwich, shows the flagship of Admiral Nelson leading the Windward Column of 12 battleships in line astern into the engagement with the *Bucentaure* (flagship of the French Navy, Admiral Villeneuve): what is particularly fascinating in this painting is that it shows the precise moment at which the foretopsail was brought down onto the deck of the *Victory* early in the battle, probably by gunfire from the French battleship *Neptune.* The Leeward Column of 15 Royal Navy battleships under the command of Vice-Admiral Collingwood engaged with the Spanish fleet under the command of Admiral Gravina in the *Santa Ana*.

Resins have been used as coloring agents and were adopted as early pharmaceuticals by many cultures; a good example is that of *dragon's blood*, a resin that came into prominence in the Roman period. A particularly valued extract was originally harvested from the fruit of the dragon trees in Socotra, East Africa, but other varieties are now known globally. As for the ambers and copals described above, the establishment of a molecular database of specimens of resin from designated geographical locations provides a means whereby analytical science can advise museum curators which botanical specimen they are actually dealing with for restoration purposes and its geographical source; hence, examples of *Dracaena, Daemonorops,* and *Croton* resins, which are all generically referred to as *dragon's blood*, have been characterized molecularly and analytically to provide sourcing information about their use on archaeological and art artifacts of cultural heritage importance [40, 41].

Finally, a biomaterial that has historically been used by many different cultures over many hundreds of years is ivory: a generic name for the exoskeletal teeth and other teeth of terrestrial and marine mammals—including the African elephant, Asian elephant, sperm whale, narwhal, hippopotamus, walrus, and warthog. Other ivory substitutes have also been incorporated into this category over the ages, and modern fakes are quite prevalent, being normally composed of resin simulations with mineral additives to increase their weight and provide a suitable textural balance. Forgeries also exist of sperm whale teeth which have been carved with marine scenes and legends by sailors on whaling ships in the 18th and 19th centuries; these are meant to deceive the incautious purchaser; in this respect, the whole artifact may be faked and comprise a resin simulation with applied carving, or in some instances, the ivory is genuine but the carving is a forgery that has been applied to enhance its resale value. For some years, analytical spectroscopy has been used to determine whether or not a piece of carved ivory is genuine or a simulation, and it has been found that genuine examples in museum collections are often interspersed with fakes. The animal origin of the ivory artifact is also important information from an archaeological point of view, as it informs historians about trade routes, and molecular spectral databases exist to determine this: as well as for the purpose of advising museum curators, this information is important for the CITES law enforcement teams who operate in many countries to prevent the carriage and trafficking of endangered species, including animal skins, teeth, etc. and several papers exist in the forensics literature which reflect the advancement in this particular application. As with resins, the establishment of a molecular database is essential for comparison purposes and this facilitates an analytical discrimination to be made between terrestrial elephant, hippopotamus, and warthog ivories and their marine analogs of the sperm whale, walrus, and narwhal [42, 43]. A rather special analytical exemplar is provided by mammoth ivory, which is occasionally found in rather large quantities in the permafrost and ancient gravel beds; specimens can date back to 60 Kya or more. Whereas it is generally possible to analytically distinguish between pieces of ancient mammoth and modern elephant ivory, the degradation that may have been suffered by the mammoth ivory in its depositional environment is very dependent upon the particular environmental conditions to which it has been exposed, such as the acidity of the burial soil, etc. Ivory is composed of an organic keratotic collagenic component in an inorganic hydroxyapatite phosphatic matrix (in the ratio of approximately 1:4), and the organic component is first degraded and then leached out with time, being replaced with a silicaceous component in the channels and voids left in the mineralized phosphatic matrix. It has been estimated that up to 20% of the ivory being illegally traded globally currently is mammoth ivory, so its analytic identification is a prime objective in discriminating between ancient ivory and its modern analogs. Despite claims that have been made to the contrary, it is really impossible to measure the age of archaeological ivory from the intensity ratios of the component bands of the organic keratin and inorganic phosphatic matrix in the molecular spectra because of the indefinite nature of the environmental degradation that has ensued during its burial [29], which selectively destroys the organic keratotic signatures.

1.6 A survey of the chapters contained in this volume

As part of this new introductory chapter to this second edition, a survey of the analytical techniques described in the contributing chapters is necessary. The nine chapters which comprise this second edition can be subdivided into four chapters taken from the first edition, three new chapters, one updated chapter, and this new introductory chapter, which for the first time surveys the wider range of analytical techniques, predominantly elemental and molecular in nature, that have historically been utilized in the analytical interrogation of cultural heritage artifacts. The coverage of these chapters can be described as follows:

- Spectroscopy as a nondestructive technique in identifying cultural archaeological heritage. From the first edition, but now includes three new case studies which describe the curation of archaeological relics that have been subjected to noninvasive investigation taken from papers delivered between 2000 and 2018 on the topics of mobile instrumentation for archaeological fieldwork, fossilized materials, and ion-beam studies of archaeological and art artifacts.
- The ESR dating application for the chronometric analysis of ancient geological materials up to 2.6 M years old based upon measuring the exposure of

such materials to natural ionizing radiation and the number of paramagnetic centers that have accumulated over time in specimens. A broad range of materials is covered, including stalactites, quartz crystals, mollusc shells, corals, speleotherms, and fossilized teeth. This chapter is essentially unchanged from its appearance in the first edition.

- The development of low-frequency EPR instrumentation for the study of artifacts of cultural heritage significance. Such instrumentation operates within the frequency range of 100–500 MHz. Although the application of the high-frequency EPR technique, which operates at 9 GHz, is well established, a big disadvantage has been the limited access to specimens of rather small size (typically of a few mm^2 in section), which is critically dependent upon their containment in the sample housing of the instrument—which leads to the technique being rejected as noninvasive, because sampling is required. The low-frequency adaptation permits the study of larger specimens, and several applications are described. This is also a chapter that has been taken essentially unchanged from the first edition.
- The ESR determination of the temperature of prehistoric lithic heat treatments: this is widely used to obtain information about the processes that a sample has undergone geologically and synthetically; it concerns the thermal history of minerals and their interconversion. It is especially useful when applied to lithic tools that were subjected to some form of heat treatment in antiquity. Again, this is a chapter that has been taken from the first edition.
- The fourth chapter, which has been taken unchanged from the first edition, describes the ESR determination of the age of the Eastern Romanian Plain loess in combination with the TL technique. The paramagnetic centers in the quartz crystals in the specimens were subjected to both ESR and TL analyses, and the geological age was found to be in good agreement with values of 19 ± 2 and 20 ± 2 K years old, respectively, obtained through TL and ESR measurements.
- The first of three new chapter entries for this second edition, in addition to this introductory survey chapter, describes the ESR dating of mollusc shells, corals, and other materials, forming a study of their paramagnetic electron traps. The limit of detection currently is around 500 Kya, and the chapter recounts details of the instrumentation and sample preparation required.
- The second of three new chapters for this second edition describes nonlinear optical microscopic techniques, which essentially provide imaging information stratigraphically and nondestructively for a variety of cultural heritage materials of biological and inorganic origins. Depth profiling reveals details about the chemical and physical composition of layers in artworks, varnishes, glues and finishing layers, parchment, and corrosion layers on silver substrates.
- The final of these four new chapters incorporated into this second edition describes the application of portable and nondestructive XRF spectroscopy to pottery and pigments associated with cultural heritage artifacts. A wide range of elements can be detected in major and minor quantities; this assists

in the identification of the chemical composition of the artifact and can in some instances be material in ascribing a source to the raw materials based on the presence of the minor elements detected. As the element detectability is dependent upon the square root of its atomic number, the lighter elements such as boron cannot be seen; however, elements with an atomic number >20 can normally be detected using this technique. The portability of the instrumentation means that it can easily be used in a museum environment and particularly where there are restrictions imposed by the size of the artifact.

1.7 Summary

A summary of the chapters that follow and are included in this second edition can be made on the basis that they are all based on elemental detection, whether this be for dating or chemical composition purposes. The exceptions are the new chapter on nonlinear optical microscopy, which relies upon imaging techniques rather than the presence of a distinctive elemental verification of composition, and this current introductory chapter which gives an overall appreciation of many more elemental and molecular techniques that have been used for cultural heritage artifacts and materials and which provide a perspective for the general reader. In the latter case of molecular techniques, the applications have generally been well covered by several texts (see, for example, table 1.2) that do not in general impinge upon the complementary physical techniques that are described here.

A clear take-home message regarding the future analytical interrogation of cultural heritage artifacts and their raw materials is that the analyses are preferably if not essentially to be conducted in a nondestructive or noninvasive mode, using portable instrumentation which negates the need for chemical or mechanical pretreatment of the specimen or its removal from secure storage and containment in a museum environment. The author of this chapter has personally experienced a secondary aspect of his analytical work on cultural heritage artifacts, namely that portable analytical instrumentation has also been used as a first-pass screening device to identify particular areas of interest that would be well served by a subsequent thorough investigation using laboratory instrumentation using micro-sampled specimens and complementary techniques, some of which are not yet available for on-site usage in a portable instrumental form. Museum conservators are usually in tune with this aspect of sampling and procedure, as the analytical benefits accruing from such microsampling can then be quantified more readily, especially when the laboratory-based studies on the microspecimens can then involve additional instrumentation that is neither portable nor nondestructive, thus giving a greater potential reward in terms of chemical and physical data for a relatively insignificant additional outlay.

References

[1] Ferchault de Réaumur R-A 1727 Idée générale des différentes manières dont on peut faire de la porcelaine et quelles sont les véritables matières de celle de la Chine *Mem. Acad. Sci. Paris*

185–203 https://www.academie-sciences.fr/pdf/dossiers/Reaumur/Reaumur_pdf/p185_203_vol3589m.pdf

[2] Ferchault de Réaumur R-A 1729 Second mémoire sur la porcelaine ou suite des principes qui doivent conduire dans la composition des porcelaines de différents genres et qui établissent les caractères des matières fondantes qu'on ne peut choisir pour tenir lieu de celle qu'on emploie à la Chine *Mem. Acad. Sci. Paris* 325–344 https://www.academie-sciences.fr/pdf/dossiers/Reaumur/Reaumur_pdf/p325_344_vol3527m.pdf

[3] Ferchault de Réaumur R-A 1739 Memoire sur l'art de faire une nouvelle espace de porcelana pardes moyens extremement simples et fáciles ou de transformer le verre en porcelaine *Mem. Acad. Sci. Paris* 370–388 https://www.academie-sciences.fr/pdf/dossiers/Reaumur/Reaumur_pdf/p370_388_vol3536m.pdf

[4] Davy S H 1815 Some experiments and observations on the colours used by the ancients *Phil. Trans. R. Soc.* **105** 97–124

[5] Eccles H R and Rackham B 1922 *Analysed Specimens of English Porcelain in the Victoria and Albert Museum Collection* (London: Victoria & Albert Museum Publications)

[6] Church S A H 1885 and 1894 *English Porcelain: A Handbook to the China Made in England during the 18th Century as Illustrated by Specimens Chiefly in the National Collection* (London: Chapman and Hall)

[7] Soden-Smith R 1873 *Catalogue of the Collection of English Pottery and Porcelain Exhibited on Loan at the Alexandra Palace, 1873* (London: RK Burt & Co.)

[8] Edwards H G M 2020 *18th and 19th Century Porcelain Analysis: A Forensic Provenancing Assessment* (Dordrecht: Springer-Nature)

[9] Edwards H G M 2022 *Porcelain Analysis and Its Role in the Forensic Attribution of Ceramic Specimens* (Dordrecht: Springer-Nature)

[10] Burgio L, Clark R J H and Hark R H 2009 Spectrosocpic investigation of modern pigments on purportedly medieval miniatures by the Spanish Forger *J. Raman Spectrosc.* **40** 2031–36

[11] Coomans T and de Maeyer J 2007 The Spanish forger: master of manuscript chicanery *The Revival of Medieval Illumination: Nineteenth Century Belgian Manuscripts and Illumination from a European Perspective* (Leuven: Leuven University Press)

[12] Voelkle W and Wiecke R 1979 *The Spanish Forger* (New York: Pierpont Morgan Library)

[13] Orna M V 1996 Chemistry at the interface of archaeology and art *Chem. Aust.* 470

[14] Craddock P 2009 *Scientific Investigation of Copies, Fakes and Forgeries* (Oxford: Butterworth-Heinemann/Elsevier Press)

[15] Clark R J H and Gibbs P J 1997 Identification of lead(II) sulfide and pararealgar on a 13th century manuscript by Raman microscopy *Chem. Commun.* 1003–4

[16] Edwards H G M, Domenech-Carbo M T, Hargreaves M D and Domenech A 2008 A Raman spectroscopic and combined analytical approach to the restoration of severely damaged frescoes: the Palomino project *J. Raman Spectrosc.* **39** 444–52

[17] Edwards H G M, Farwell D W, Seaward M R D and Giacobini C 1991 Preliminary Raman microscopic analysis of a lichen encrustation involved in the biodeterioration of Renaissance frescoes in Central Italy *Int. Biodeterior.* **27** 1–9

[18] Edwards H G M and Seaward M R D 1993 Raman Spectroscopy and lichen biodeterioration *Spectrosc. Eur.* **5** 16–20

[19] Brown K L and Clark R J H 2002 Analysis of pigmentary materials on the Vinland Map and Tartar Relation by Raman microprobe spectroscopy *Anal. Chem.* **74** 3658–61

[20] Chalmers J M, Edwards H G M and Hargreaves M D 2012 *Infrared and Raman Spectroscopy in Forensic Science* (Wiley: Chichester)

[21] Edwards H G M 2012 The Vinland Map: an authentic relic of early exploration or a modern forgery—Raman spectroscopy in a pivotal role? *Infrared and Raman Spectroscopy in Forensic Science* ed J M Chalmers, H G M Edwards and M D Hargreaves (Chichester: Wiley) 7b 401–7

[22] Edwards H G M and Chalmers J M 2005 *Raman Spectroscopy in Archaeology and Art History* (Cambridge: Royal Society of Chemistry)

[23] Edwards H G M and Lewis I R 2001 *Handbook of Raman Spectroscopy: From the Research Laboratory to the Process Line* (Practical Spectroscopy Series 28) (New York: Marcel Dekker)

[24] Edwards H G M 1999 Art works studied by infrared and Raman spectroscopy *Encyclopaedia of Spectroscopy and Spectrometry* ed J C Lindon, G Tranter and J L Holmes (London: Academic) pp 2–17

[25] Sgamellotti A, Brunetti B G and Miliani C 2014 *Science and Art: The Painted Surface* (Cambridge: Royal Society of Chemistry)

[26] Edwards H G M, Vandenabeele V and Colomban P 2023 *Raman Spectroscopy in Cultural Heritage Preservation* (Dordrecht: Springer)

[27] Ciliberto E and Spoto G 2000 *Modern Analytical Methods in Art and Archaeology* (Wiley: Chichester)

[28] Pollard A M and Heron C 1996 *Archaeological Chemistry* (Cambridge: Royal Society of Chemistry)

[29] Vandenabeele P and Edwards H G M 2018 *Raman Spectroscopy in Archaeology and Art History* vol II (Cambridge: Royal Society of Chemistry)

[30] Edwards H G M 2004 ArchaeoRaman: probing history with Raman spectroscopy *Analyst* **129** 870–79

[31] Edwards H G M, Farwell D W, Rull Perez F and Jorge Villar S E 1999 Spanish mediaeval frescoes at Basconcillos del Tozo: an FT Raman spectroscopic study *J. Raman Spectrosc.* **30** 307–12

[32] De Faria D L A, Edwards H G M, Afonso M C, Brody R H and Morais J L 2004 Raman spectroscopic analysis of a *tembeta:* a resin archaeological artefact in need of conservation *Spectrochim. Acta A* **60** 1505–13

[33] Brody R H, Edwards H G M and Pollard A M 2001 A study of amber and copal resins using Fourier transform Raman spectroscopy *Spectrochim. Acta A* **57** 1325–38

[34] Vandenabeele P, Grimaldi D V, Edwards H G M and Moens L 2003 Raman spectroscopic study of different types of Mexican copals and related resins *Spectrochim. Acta A* **59** 2221–30

[35] Spindler K 1993 *The Man in the Ice* (London: Wiedenfeld & Nicholson)

[36] Williams A C, Edwards H G M and Barry B W 1995 The iceman: molecular structure of 5200 years old skin characterised by Raman spectroscopy and electron microscopy *Biochim. Biophys. Acta* **1246** 98–105

[37] Edwards H G M, Williams A C, Farwell D W, Barry B W and Rull F 1995 Novel deconvolution procedure for complex biological systems: vibrational components in the Fourier transform Raman spectrum of iceman and contemporary skin *J. Chem. Soc. Faraday Trans.* **91** 3883–87

[38] Edwards H G M and Falk M J 1995 Investigation of the degradation products of archaeological linens by Raman spectroscopy *Appl. Spectrosc.* **51** 1134–38

[39] Edwards H G M, Wyeth P, Farwell D W, Nikhassan N F and Garside P 2006 Raman spectroscopic analysis of an important archival linen specimen: the foretopsail of HMS Victory *J. Raman Spectrosc.* **37** 1193–200
[40] Edwards H G M, Farwell D W and Quye A 1997 Dragon's blood I: characterisation of an ancient resin using Fourier transform Raman spectroscopy *J. Raman Spectrosc.* **28** 243–49
[41] Edwards H G M, de Oliveira L F C and Prendergast H V Raman spectroscopic analysis of dragon's blood resins—a basis for distinguishing Dracaena (Convallariaceae), Daemonorops (Palmae) and Croton (Euphorbiacea) *Analyst* **129** 134–38
[42] Edwards H G M and Farwell D W 1995 Ivory and simulated ivory artefacts: an FT Raman diagnostic study *Spectrochim. Acta A* **51** 2073–82
[43] Edwards H G M, Brody R H, Nikhassan N F, Farwell D W and O'Connor S 2006 Identification of archaeological ivories using Fourier transform Raman spectroscopy *Anal. Chim. Acta* **559** 64–72

Chapter 2

Nonlinear optical microscopy techniques for cultural heritage studies

Meropi Mari and George Filippidis

Nonlinear optical microscopy modalities are well-established techniques for biological applications. In the last fifteen years, they have been used as noninvasive, label-free, high-resolution diagnostic tools for the in-depth, stratigraphic analysis of various cultural heritage (CH) objects of high significance. These imaging techniques render complementary information about the chemical composition and physical structure of materials such as fresh and aged protective varnish layers, model painted artworks, lining glues, historical coatings, parchments, and the corrosion layers of silver-based artifacts.

2.1 Introduction

A nonlinear optical process occurs when 'the response of a material system to an applied optical field depends in a nonlinear manner upon the strength of the optical field' [1]. The first described nonlinear optical phenomenon, the quantum mechanism of two-photon molecular excitation, was presented by Maria Goeppert Mayer in her doctoral thesis in 1931. Lasers are light sources that have sufficient intensity to produce nonlinear optical phenomena, while, at the same time, a large number of laser applications owe their existence to nonlinear optics [2]. In recent years, nonlinear optical microscopy (NLOM) has become a well-established microscopy technique, because femtosecond pulsed Ti:sapphire lasers, which have the intensity required to induce adequately detectable nonlinear signals, have became commercially available; these are compact, user-friendly systems [3]. Nonlinear optical imaging techniques based on the employment of ultrafast near-infrared lasers as excitation sources have been widely used to reveal biological structures for accurate diagnosis at the cellular level as well as the tissue level [4–6]. This nondestructive

doi:10.1088/978-0-7503-4846-1ch2

technology offers several advantages compared to other optical techniques (such as confocal microscopy). Specifically, due to the use of excitation wavelengths located in the near-infrared region of the spectrum, where the absorption and scattering phenomena from CH materials are reduced, NLOM enables information to be extracted from deeper layers. In addition, due to the nonlinear nature of this procedure, NLOM provides intrinsic 3D optical sectioning, images at increased axial and lateral resolutions, and prolonged irradiation periods. Only a small volume of laser light interacts with the sample, thus minimizing thermal deposition or other photodamage effects [7]. These characteristics are requisite for CH studies. Moreover, second- and third-harmonic generation (SHG, THG) modalities are intrinsic properties of the material. Thus, there is no need for sample preparation, indicating that these nonlinear scattering modalities are label-free imaging techniques. The simultaneous acquisition of different nonlinear signals, such as multiphoton excitation fluorescence (MPEF), SHG, and THG provides complementary information not only about structure and morphology but also about the composition of specimens. The above unique characteristics make NLOM an innovative, minimally invasive diagnostic imaging tool for CH studies.

Artworks such as paintings consist of a variety of layers with different chemical compositions [8]. In short, there is a 'support' layer, which is the base of the painting, and a 'ground' layer that is between the paint and support layer. The support layers are usually covered by various glues to ensure sufficient adhesion. A painted artwork also consists of the paint layer, underdrawings, and underpaintings. The 'paint' layer is a mixture of the pigment (fine powder) and a binder (a substance that keeps the powder uniform). The outer layer is a 'varnish' layer, which is a transparent coating that provides protection to the artwork and/or acts as a glazing agent [9, 10].

In recent decades, a significant effort has been made to develop novel, minimally invasive techniques in the field of CH science. A number of optical techniques, ranging from micro-destructive ones in which sampling is necessary to noninvasive ones, have been employed as diagnostic tools for artworks. Specifically, optical coherence tomography (OCT) [11], Raman spectroscopy [12], Fourier-transform infrared spectroscopy (FTIR) [13], laser-induced breakdown spectroscopy (LIBS) [14], holography [15], photoacoustic microscopy [16], multispectral imaging [17], laser-induced fluorescence spectroscopy (LIF) [18] and terahertz time-domain imaging (THz-TDI) [19] are some representative techniques based on electromagnetic radiation that have been used for CH studies.

Recently, NLOM was tested for the noninvasive in-depth analysis of various CH objects using the modalities of MPEF [20], SHG [21] and THG [22]. It was shown that these modalities can provide new insights in the stratigraphic analysis of complex CH objects and extract information about the layered structure/morphology of the samples. It was also shown that these modalities have the potential to reveal useful information about the chemical composition and the physical structure of the studied CH objects. Thus, they can help art historians and conservators to reconstruct the history of the sample, evaluate its authenticity, obtain information about the artist's technique, and provide valuable support to restorers to help them to accurately control any cleaning interventions. In the next section, we discuss one-photon excitation fluorescence (OPEF); we subsequently describe the physical principles of NLOM with a focus on nonlinear absorption (two-photon excitation fluorescence, which is the simplest case of MPEF); finally, the nonlinear coherent scattering phenomena of SHG and THG are presented.

2.2 Imaging techniques

In OPEF, an electron from the ground-state level of a molecule interacts with a photon. If the photon energy is higher than the energy difference between the first excited electronic state and the ground state, then the photon is absorbed by the electron, thus causing the electron to transition to a first excited-state level. Subsequently, through non-radiative processes such as thermal relaxation, the excited electron loses a part of its energy and transitions to the lowest vibrational level of the excited state. Finally, the radiative electron deactivates and returns to the ground state via the spontaneous emission of a photon; this phenomenon is called fluorescence (figure 2.1(A)). The absorption process lasts only a few femtoseconds, while the fluorescence process takes place in nanoseconds.

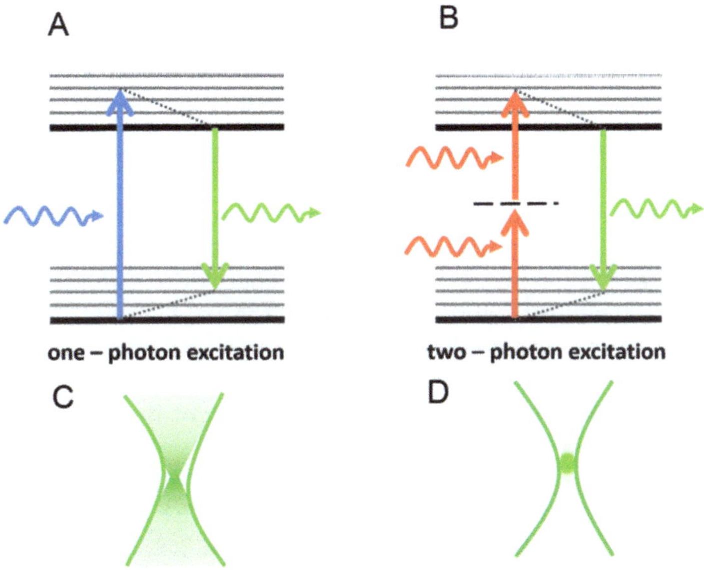

Figure 2.1. The upper images depict the Jablonski diagrams for (A) one- (OPEF) and (B) two-photon excitation fluorescence (TPEF). The horizontal lines denote the energy eigenvalues of the molecule, where the black and the gray lines correspond to the electronic and vibrational states, respectively. In addition, the blue or red and green vertical lines denote the absorbed and emitted photons, respectively. The dotted lines represent non-radiative deactivation due to the thermal relaxation of the electron. Panel (D) illustrates the localization of the two-photon excitation volume. TPEF occurs at a localized 3D spot. It can be calculated that for TPEF, the focal volume is ~2 μm^3 (for an excitation wavelength of 1000 nm and a numerical aperture (NA) 0.8 objective lens). Thus, the excitation region is limited to the focal plane for the two-photon case, while it extends above and below the focal plane for one-photon excitation (C).

The emitted photon has a longer wavelength compared to the absorbed photon, since part of its energy is converted to heat or kinetic energy of the system due to the aforementioned thermal relaxation of the electron. This effect is known as the Stokes shift and facilitates the separation of the fluorescent signal from the excitation radiation. As shown by equation (2.1), the OPEF absorption rate (in photons/s) of a molecule W_1 increases linearly with the incident photon flux, F (photons $cm^{-2}s^{-1}$), and the photon flux is proportional to the intensity I (*$I=F{\cdot}\hbar{\cdot}\omega$, $\hbar$: reduced Planck's constant, ω: frequency of incident light*). It is obvious that OPEF is a linear phenomenon. The variable σ_1 of equation (2.1) is the cross section of the OPEF and typically has a value on the order of 10^{-17} cm^2.

$$W_1 = \sigma_1 \cdot F \tag{2.1}$$

On the other hand, MPEF is a nonlinear phenomenon. The simplest case of MPEF is TPEF. In this case, two photons, located in the infrared region of the spectrum, interact with an electron and produce one fluorescence photon. Specifically, an electron from the ground state of a molecule interacts with a photon whose energy is slightly larger than half of the energy difference between the first excited electronic state and the ground state. The electron is then excited to an intermediate energy level between the ground state and the first excited state, generally known as a 'virtual state.' However, the virtual state is not an eigenstate of the molecule; thus, the lifetime of an electron in this state is extremely short. A good approximation that can be used to estimate the electron's lifetime Δt in the virtual state is the uncertainty of energy ΔE ($\Delta E{\times}\Delta t \approx \hbar$). If we consider that ΔE is about half the energy difference between the first excited state and the ground state and has a typical value of 1 eV, then the electron's lifetime in this virtual state is about 0.5 fs. If, during this short time, the electron interacts with a second identical photon, it will be excited to a vibrational level of the first excited molecule state. As in the OPEF case, the electron loses part of its energy through thermal relaxation and transitions to the lowest vibrational level of the excited state. The radiative de-excitation of the electron to a vibrational or a rotational level of the ground state via the spontaneous emission of a photon is called fluorescence. In this case, the energy of the emitted photon is smaller than the sum of the two incident infrared photons (figure 2.1(B)).

However, the probability of the electron interacting with a second photon while it is in the virtual state is extremely low; for this reason, TPEF needs a high photon flux in order to occur. As mentioned above, the photon flux is proportional to the intensity of the incident beam, which is proportional to the power of the source and inversely proportional to the irradiated area. Short-pulse-duration lasers (femto-second) and high numerical aperture (NA) lenses are required to generate this phenomenon. High NA lenses ensure that the irradiation is tightly focused in a small area, while ultrashort-pulse lasers produce the instantaneous high peak power that is required for the production of TPEF signals. The absorption rate W_2 of TPEF, as shown by equation (2.2), confirms the nonlinear (quadratic) dependence of the phenomenon on the incident beam flux F and therefore on the intensity.

$$W_2 = \frac{1}{2} \cdot \sigma_2 \cdot F^2 \tag{2.2}$$

The variable σ_2 represents the cross section of the phenomenon, which has a typical value on the order of 10^{-50} $cm^4 \cdot s$ per photon. The cross section of TPEF, σ_2, is much smaller than the cross section of OPEF. The combination of this low cross-sectional value, the nonlinear dependence, and the irradiation intensity give NLOM one of its most important advantages: high 3D resolution.

As figure 2.1(D) depicts, the TPEF signals occur at a tiny spot in the center of the focused beam. Within this volume, the intensity is high enough to produce detectable TPEF signals. Due to the nonlinear dependence of the phenomenon on the incident intensity, the size of the excited spot is extremely small. The size of the excited spot determines the resolution of the imaging technique. As a consequence, the resolution of NLOM starts at a few μm and drops to a few hundreds of nm, particularly in the lateral direction. At the same time, side effects such as photodamage and photobleaching are minimized in all regions other than the limited focal spot.

SHG and THG are coherent, nonlinear scattering phenomena. Specifically, two or three photons of angular frequency ω are destroyed and a photon of angular frequency 2ω (for SHG) or 3ω (for THG) is simultaneously created in a single quantum-mechanical process (figure 2.2). Due to its virtual level-transition characteristic, higher harmonic generation does not deposit any energy in the matter thus interacted with, since there is no electronic transition involved and thus the emitted photon energy is exactly the same as the total absorbed photons' energies. These virtual-transition and energy-conservation characteristics provide the optical 'non-invasive' nature desirable for CH studies. Moreover, SHG and THG are intrinsic properties of the material. Thus, there is no need for sample preparation or staining to generate these signals, indicating that these modalities are label-free imaging techniques.

High SHG signals arise from non-centrosymmetrically oriented structures. The most common CH materials that produce detectable SHG signals are collagen and starch granules. In contrast to SHG, THG can occur in any medium. However, increased THG signals arise from optical inhomogeneities, such as interfaces between two media, of comparable size to the beam focus. Thus, this modality is

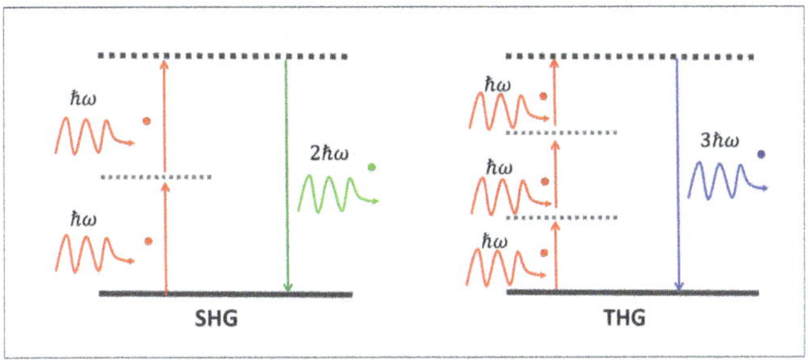

Figure 2.2. Energy level diagrams describing SHG (on the left) and THG (on the right).

ideal for probing refractive index changes and providing submicrometer structural and morphological information.

2.3 Experimental setup

A characteristic experimental apparatus with two pathways of detection, one in the backward and another in the forward direction, is depicted in figure 2.3. Tightly focused femtosecond laser pulses represent ideal sources for nonlinear microscopy applications.

A femtosecond laser source is required for the excitation of the samples. This femtosecond pulsed laser emits high enough instantaneous intensities to produce detectable nonlinear signals. In addition, high NA objective lenses are employed to achieve tight focusing of the laser beam onto the specimen. The power at the focal plane can be adjusted by the use of intensity filters in front of the laser output. The raster scanning procedure (in the lateral *xy*-direction) is performed by a pair of galvanometric mirrors. A high degree of linear laser polarization is achieved through the use of a linear polarizer. In some cases, a phase retardation wave plate can be employed in order to rotate the laser polarization to a specific angle. A telescope system can be used, which consists of two converging lenses that expand the diameter of the laser beam in order to fill the back aperture of the objective lens. The selection of the proper focal plane is achieved by employing a motorized *xyz* translation stage. The sample is usually mounted on this stage using a base. Nonlinear signals are collected in the backward direction by using a photomultiplier tube (PMT). Appropriate filters are placed in front of the PMT to eliminate the reflected laser light and solely detect the nonlinear signals arising from the irradiated CH samples. Signals can also be detected in the forward direction by employing a condenser lens, a second PMT, and appropriate filters for the spectral filtering. Moreover, the setup provides a module that is able view to the sample using an illumination source (a white lamp) and observe it through a charge-coupled device (CCD) sensor.

The resolution of this apparatus depends on the effective volume of the spot, which is defined as the volume of the focused beam within which there is sufficient

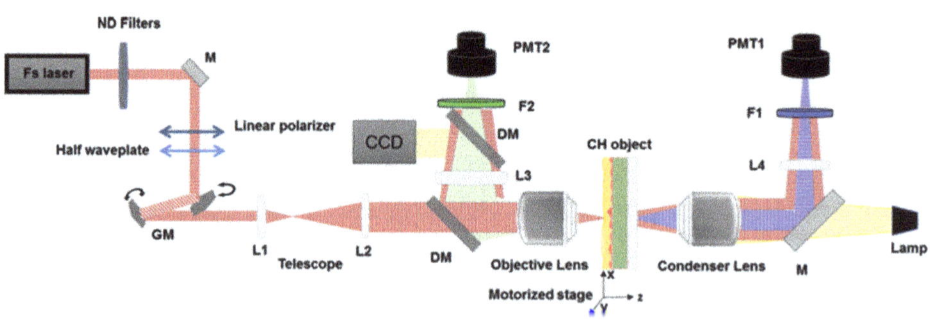

Figure 2.3. A characteristic experimental apparatus. fs laser: femtosecond laser source, ND filters: neutral density filters, M: mirror, GM: galvanometric mirrors, L: lenses, DM: dichroic mirror, PMT: photomultiplier tube, F: filters, CCD: charge-coupled device.

intensity to produce the nonlinear signal. If no aberrations are taken into account, the radius r of the spot at the focal plane equals [23]:

$$r = 0.61\frac{\lambda}{NA}, \tag{2.3}$$

where λ is the wavelength of the laser beam and NA is the numerical aperture of the objective lens. It should be noted that in the case of nonlinear induced phenomena, the lateral resolution R is reduced to:

$$R = \frac{r}{\sqrt{m}}, \tag{2.4}$$

where m denotes the degree of nonlinearity of the power dependence of the produced signal on the incident beam. For example, in the case of THG, $m=3$, while for SHG and TPEF, $m=2$.

Following the same approach as that used for the lateral resolution, the diffraction-limited axial resolution of NLOM setup can be calculated through the next equation:

$$R_{\text{axial}} = \frac{n{\cdot}\lambda}{NA^2}{\cdot}\frac{1}{\sqrt{m}}, \tag{2.5}$$

where n is the refractive index of the material into which the laser beam is focused.

Characteristic values for the lateral and axial resolutions of a system are ~500 nm and ~2 μm, respectively.

2.4 The applications of nonlinear optical microscopy to cultural heritage objects

The use of innovative, noninvasive, optical diagnostic tools that have the potential to provide new insights into the assessment of the appropriate conservation method for various CH objects can be extremely useful to restoration scientists. The first demonstration of NLOM on CH objects involved the submicron thickness identification of different natural and synthetic protective varnish layers via THG and the discrimination of their chemical compositions through MPEF measurements [22]. This has been demonstrated in both the transmission mode and the reflection mode [24]. Previous studies have also demonstrated that MPEF signals can distinguish between fresh and aged varnish samples [25]. In addition, SHG and polarization-dependent SHG (PSHG) microscopies are able to provide reliable information about the compositions of various types of natural and synthetic adhesive used in the conservation or restoration of painted artworks or papers and books of cultural significance and for the quantitative discrimination of aged from fresh starch-based glues, respectively [21, 26]. Another study has confirmed the feasibility of NLOM as a means of detecting the corrosion layer of silver-based artifacts at high resolution [27]. Moreover, these nonlinear imaging techniques have been used to determine the extent of the photochemical damage that could be caused to underlying paint layers during the laser removal of protective varnish coatings [28, 29]. This diagnostic approach can lead to the identification of the optimal laser cleaning conditions that cause the minimum collateral damage to the paint layers. MPEF has been applied for the evaluation of phthalocyanine acrylic paint layer thickness [30], to probe the stratigraphy of egg tempera mock-up paintings [31], and to characterize grisaille paint layers on historical stained glasses [32]. Recently, the same nonlinear modality has been employed to determine the thickness of model medieval-type glass layers in a nondestructive manner [33]. Furthermore, nonlinear imaging techniques have been applied in order to extract information from historical coating samples [34]. Recent works have also demonstrated that the degradation of parchment can be revealed by a combination of SHG and PSHG intensity signals [35, 36].

The next section analytically discusses some characteristic applications of all the above nonlinear microscopy modalities to CH objects. Specifically, we present the application of THG and MPEF measurements to a varnish sample that is employed as a protective layer in an artifact. In addition, SHG imaging is used to monitor the collagen distribution in a parchment sample.

Natural varnish layers are the materials most widely used to protect painted surfaces from atmospheric pollution and oxidation and to improve the aesthetic appearance of artworks. However, both time and exposure to atmospheric conditions may irreversibly modify varnish layers, thus requiring their removal. The removal of the surface varnish layer comprises a delicate intervention in CH conservation. Nonlinear imaging is a high-resolution (on the order of μm), nondestructive technology that provides the precise determination of the protective varnish layer thickness. This information can facilitate the cleaning of artworks.

Figure 2.4 presents THG (a) and three-photon excitation fluorescence (3PEF) (b) imaging used to determine the thickness of a varnish layer (dammar) laid on a glass coverslip. Dammar is a natural resin commonly used as a surface coating in the conservation of artworks. The dammar was dissolved in turpentine, creating a

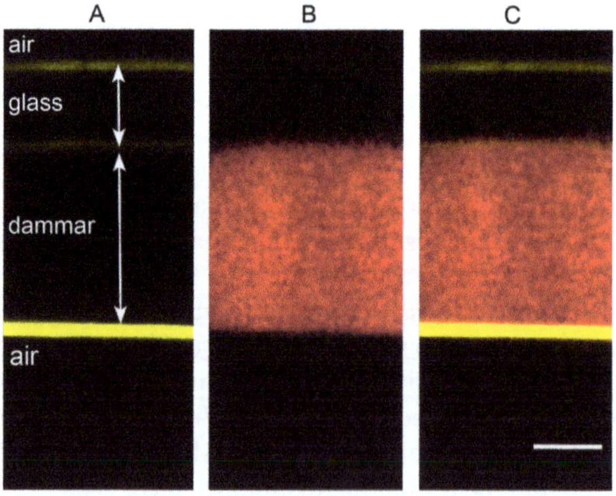

Figure 2.4. Cross-sectional image of a dammar sample: (A) THG, (B) MPEF, (C) multimodal image (THG yellow, MPEF red). The scale bar depicts 50 μm.

transparent layer. The solution was cast onto a thin round glass microscope slide. The 3PEF and THG signals were collected simultaneously in backward (3PEF) and transmission (THG) modes. Ultraviolet–visible (UV–vis) absorption analysis verified the transparency of dammar in the near-infrared (NIR) and visible parts of the spectrum and maximum absorption of ultraviolet (UV) wavelengths. These results confirmed that the fluorescence observed after the irradiation of the dammar sample with the 1028 nm fs laser was a three-photon excitation process. Concerning the acquisition of THG in transmission mode in a multilayer sample, the refractive index mismatch between the media allowed the different interfaces to be distinguished. Specifically, using THG measurements, three different layers can be distinguished, corresponding to the interfaces between the different media (figure 2.4(A)). The upper interface corresponds to the air/glass interface and the lower one to the dammar/air interface. The middle layer represents the dammar/glass interface. Moreover, 3PEF signals arising only from the dammar layer (figure 2.4(B)) provided complementary information in a single measurement, which confirmed that the material thickness was 201 μm. Figure 2.4(C) is a multimodal image of the fresh dammar sample obtained via THG and MPEF.

Figure 2.5 shows an SHG image of a parchment (calf) sample. Parchments are among the most sensitive materials in a museum's collection. Information related to their composition can support their restoration procedures. A main ingredient of these samples is collagen. Collagen is an emitter of high SHG signals due to its highly crystalline, non-centrosymmetric triple-helix structure [5]. Thus, SHG imaging is an ideal diagnostic tool that can monitor collagen fiber distribution and its discontinuities in parchments. Recently, it has been shown that analysis performed using PSHG measurements can provide information for the *in situ* monitoring of the degradation effects of parchments [36].

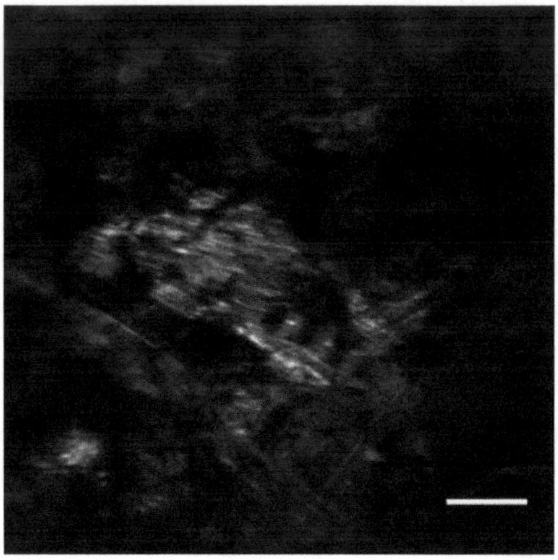

Figure 2.5. SHG image from a parchment made from calf skin. White denotes the maximum intensity and black the minimum intensity SHG signals. The scale bar depicts 40 μm.

The above results are a proof of the complementary information that NLOM can provide as a reliable diagnostic tool for various CH objects. Nonlinear imaging provides the compositional characterization of CH samples via MPEF, the submicron, in-depth morphological delineation via THG, and the non-centrosymmetric materials distribution via SHG [37, 38]. Moreover, this technology can act synergistically with other well-established techniques such as confocal Raman microscopy [39], photoacoustic tomography [40], optical coherence tomography [41] and terahertz time-domain spectroscopy [42], thereby obtaining additional information from the investigated CH materials. It should be noted that the identification of the safe limits for the use of the nonlinear techniques for CH studies has been also investigated. Some studies have evaluated the damage caused by near-infrared femtosecond pulsed lasers and established safe laser thresholds for a variety of model painted artworks [41, 43]. The target is to determine the exact experimental parameters for the safe irradiation of CH objects.

2.5 Conclusions and perspectives

The materials involved in CH science are numerous and heterogeneous. Thus, there is a need for the development of new, noninvasive techniques to fulfill a variety of requirements associated with CH studies. Nonlinear imaging provides unique, complementary information about the chemical composition and physical structure of the studied materials. NLOM modalities can be utilized as high-resolution, nondestructive tools for the in-depth, label-free, stratigraphic analysis of various CH objects. In the last few years, an increased number of research groups worldwide have used nonlinear modalities as diagnostic tools in the service of CH, thereby extracting valuable insights that have facilitated cleaning interventions. However, until now, all the experimental work in the NLOM field has been performed in controlled laboratory environments.

We expect that in the near future, it will be possible to extensively use nonlinear optical imaging techniques outdoors for the microscopic *in situ* analysis of a variety of CH objects. The main future target is the development of a custom-made, user-friendly, transportable, nonlinear optical microscope that can be employed by conservation scientists in museums and galleries. All of the miniaturization technologies needed for such a device are being developed for biomedical applications, and it is anticipated that they could easily be adapted to the peculiarities of CH and artwork studies. In addition, appropriate femtosecond tunable lasers have advanced technically in recent years; they now provide compact, stable, reliable, and ultrafast excitation sources [44].

References

[1] Boyd R W 2003 *Nonlinear Optics* (San Diego, CA: Academic)

[2] Garmire E 2013 Nonlinear optics in daily life *Opt. Express* **21** 30532

[3] Masters Barry R and So P 2008 *Handbook of Biomedical Nonlinear Optical Microscopy* (Oxford: Oxford University Press)

[4] Kuzmin N V, Wesseling P, Hamer P C, de W, Noske D P, Galgano G D, Mansvelder H D, Baayen J C and Groot M L 2016 Third harmonic generation imaging for fast, label-free pathology of human brain tumors *Biomed. Opt. Express* **7** 1889

[5] Tsafas V, Gavgiotaki E, Tzardi M, Tsafa E, Fotakis C, Athanassakis I and Filippidis G 2020 Polarization-dependent second-harmonic generation for collagen-based differentiation of breast cancer samples *J. Biophoton.* **13** e202000180

[6] Trachtenberg J T, Chen B E, Knott G W, Feng G, Sanes J R, Welker E and Svoboda K 2002 Long-term *in vivo* imaging of experience-dependent synaptic plasticity in adult cortex *Nature* **420** 788–94

[7] Meyer T, Schmitt M, Dietzek B and Popp J 2013 Accumulating advantages, reducing limitations: multimodal nonlinear imaging in biomedical sciences—the synergy of multiple contrast mechanisms *J. Biophoton.* **6** 887–904

[8] Carlyle L 2001 *The Artist's Assistant, Oil Painting Instruction Manuals and Handbooks in Britain 1800–1900 with Reference to Selected Eighteenth-Century Sources* (London: Archetype Publications Ltd)

[9] Eastaugh N, Walsh V, Siddall R and Chaplin T 2004 *Optical Microscopy of Historical Pigments* (Oxford: Elsevier)

[10] Stoner J H and Rushfield R (ed) 2012 *Conservation of Easel Paintings* (London: Routledge)

[11] Liang H, Cid M G, Cucu R G, Dobre G M, Podoleanu A G, Pedro J and Saunders D 2005 En-face optical coherence tomography—a novel application of non-invasive imaging to art conservation *Opt. Express* **13** 6133

[12] Caggiani M C and Colomban P 2011 Testing of Raman spectroscopy as a non-invasive tool for the investigation of glass-protected pastels *J. Raman Spectrosc.* **42** 790–8

[13] La Russa M F, Ruffolo S A, Barone G, Crisci G M, Mazzoleni P and Pezzino A 2009 The use of FTIR and micro-FTIR spectroscopy: an example of application to cultural heritage *Int. J. Spectrosc.* **2009** 1–5

[14] Gaudiuso R, Dell'Aglio M, de Pascale O, Senesi G S and de Giacomo A 2010 Laser induced breakdown spectroscopy for elemental analysis in environmental, cultural heritage and space applications: a review of methods and results *Sensors* **10** 7434–68

[15] Kosma K, Andrianakis M, Hatzigiannnakis K and Vivi T 2018 Digital holographic interferometry for cultural heritage structural diagnostics: a coherent and a low-coherence optical set-up for the study of a marquetry sample *Strain* **54** e12263

[16] Tserevelakis G J, Vrouvaki I, Siozos P, Melessanaki K, Hatzigiannakis K, Fotakis C and Zacharakis G 2017 Photoacoustic imaging reveals hidden underdrawings in paintings *Sci. Rep.* **7** 747

[17] Hedjam R and Cheriet M 2013 Historical document image restoration using multispectral imaging system *Pattern Recognit.* **46** 2297–312

[18] Fantoni R, Caneve L, Colao F, Fiorani L, Palucci A, Dell'Erba R and Fassina V 2013 Laser-induced fluorescence study of medieval frescoes by Giusto de' Menabuoi *J. Cult. Herit* **14** S59–65

[19] Seco-Martorell C, López-Domínguez V, Arauz-Garofalo G, Redo-Sanchez A, Palacios J and Tejada J 2013 Goya's artwork imaging with terahertz waves *Opt. Express* **21** 17800

[20] Mari M and Filippidis G 2020 Non-linear microscopy: a well-established technique for biological applications towards serving as a diagnostic tool for *in situ* cultural heritage studies *Sustainability* **12** 1409

[21] Psilodimitrakopoulos S, Gavgiotaki E, Melessanaki K, Tsafas V and Filippidis G 2016 Polarization second harmonic generation discriminates between fresh and aged starch-based adhesives used in cultural heritage *Microsc. Microanal.* **22** 1072–83
[22] Filippidis G, Gualda E J, Melessanaki K and Fotakis C 2008 Nonlinear imaging microscopy techniques as diagnostic tools for art conservation studies *Opt. Lett.* **33** 240–2
[23] Novotny L and Hecht B 2006 *Principles of Nano-Optics* (Cambridge: Cambridge University Press)
[24] Gualda E J, Filippidis G, Melessanaki K and Fotakis C 2009 Third-harmonic generation and multi-photon excitation fluorescence imaging microscopy techniques for online art conservation diagnosis *Appl. Spectrosc.* **63** 280–5
[25] Filippidis G, Mari M, Kelegkouri L, Philippidis A, Selimis A, Melessanaki K, Sygletou M and Fotakis C 2015 Assessment of in-depth degradation of artificially aged triterpenoid paint varnishes using nonlinear microscopy techniques *Microsc. Microanal.* **21** 510–7
[26] Filippidis G, Melessanaki K and Fotakis C 2009 Second and third harmonic generation measurements of glues used for lining textile supports of painted artworks *Anal. Bioanal. Chem.* **395** 2161–6
[27] Faraldi F, Tserevelakis G J, Filippidis G, Ingo G M, Riccucci C and Fotakis C 2013 Multi photon excitation fluorescence imaging microscopy for the precise characterization of corrosion layers in silver-based artifacts *Appl. Phys.* A **111** 177–81
[28] Vounisiou P, Selimis A, Tserevelakis G J, Melessanaki K, Pouli P, Filippidis G, Beltsios C, Georgiou S and Fotakis C 2010 The use of model probes for assessing in depth modifications induced during laser cleaning of modern paintings *Appl. Phys.* A **100** 647–52
[29] Oujja M, Psilodimitrakopoulos S, Carrasco E, Sanz M, Philippidis A, Selimis A, Pouli P, Filippidis G and Castillejo M 2017 Nonlinear imaging microscopy for assessing structural and photochemical modifications upon laser removal of dammar varnish on photosensitive substrates *Phys. Chem. Chem. Phys.* **19** 22836–43
[30] Dal Fovo A, Oujja M, Sanz M, Martínez-Hernández A, Cañamares M V, Castillejo M and Fontana R 2019 Multianalytical non-invasive characterization of phthalocyanine acrylic paints through spectroscopic and non-linear optical techniques *Spectrochim. Acta* A **208** 262–70
[31] Dal Fovo A, Sanz M and Oujja M *et al* 2020 In-depth analysis of egg-tempera paint layers by multiphoton excitation fluorescence microscopy *Sustainability* **12** 3831
[32] Oujja M, Agua F, Sanz M, Morales-Martin D, García-Heras M, Villegas M A and Castillejo M 2021 Multiphoton excitation fluorescence microscopy and spectroscopic multianalytical approach for characterization of historical glass grisailles *Talanta* **230** 122314
[33] Oujja M, Palomar T, Martínez-Weinbaum M, Martínez-Ramírez S and Castillejo M 2021 Characterization of medieval-like glass alteration layers by laser spectroscopy and nonlinear optical microscopy *Eur. Phys. J. Plus* **136** 859
[34] Latour G, Echard J-P, Didier M and Schanne-Klein M-C 2012 In situ 3D characterization of historical coatings and wood using multimodal nonlinear optical microscopy *Opt. Express* **20** 24623
[35] Latour G, Robinet L, Dazzi A, Portier F, Deniset-Besseau A and Schanne-Klein M C 2016 Correlative nonlinear optical microscopy and infrared nanoscopy reveals collagen degradation in altered parchments *Sci Rep.* **6** 26344

[36] Schmeltz M, Robinet L, Heu-Thao S, Sintès J M, Teulon C, Ducourthial G, Mahou P, Schanne-Klein M C and Latour G 2021 Noninvasive quantitative assessment of collagen degradation in parchments by polarization-resolved SHG microscopy *Sci. Adv.* **7** eabg1090
[37] Filippidis G, Tserevelakis G J, Selimis A and Fotakis C 2015 Nonlinear imaging techniques as non-destructive, high-resolution diagnostic tools for cultural heritage studies *Appl. Phys.* A **118** 417–23
[38] Mari M, Tsafas V, Melessanaki K and Filippidis G 2018 Applications of non-linear imaging microscopy techniques to cultural heritage objects *Insight: Non-Destr. Test. Cond. Monit.* **60** 663–9
[39] Nevin A, Comelli D, Osticioli I, Filippidis G, Melessanaki K, Valentini G, Cubeddu R and Fotakis C 2010 Multi-photon excitation fluorescence and third-harmonic generation microscopy measurements combined with confocal Raman microscopy for the analysis of layered samples of varnished oil films *Appl. Phys.* A **100** 599–606
[40] Tserevelakis G J, Tsafas V, Melessanaki K, Zacharakis G and Filippidis G 2019 Combined multiphoton fluorescence microscopy and photoacoustic imaging for stratigraphic analysis of paintings *Opt. Lett.* **44** 1154–7
[41] Liang H, Mari M, Cheung C S, Kogou S, Johnson P and Filippidis G 2017 Optical coherence tomography and non-linear microscopy for paintings—a study of the complementary capabilities and laser degradation effects *Opt. Express* **25** 19640–53
[42] Filippidis G, Massaouti M, Selimis A, Gualda E J, Manceau J M and Tzortzakis S 2012 Nonlinear imaging and THz diagnostic tools in the service of cultural heritage *Appl. Phys.* A **106** 257–63
[43] Dal Fovo A, Sanz M, Mattana S, Oujja M, Marchetti M, Pavone F S, Cicchi R, Fontana R and Castillejo M 2020 Safe limits for the application of nonlinear optical microscopies to cultural heritage: a new method for *in situ* assessment *Microchem. J.* **154** 104568
[44] Xu C and Wise F W 2013 Recent advances in fibre lasers for nonlinear microscopy *Nat. Photon.* **7** 875–82

Chapter 3

Spectrometry as a nondestructive technique in identifying cultural archaeological heritage: some case studies illustrating 'characterization without devastation'

Eman Osman

Archaeology is the exploration of humanity through the study of human and cloth relics, artifacts, and ruins. It is the primary source of our knowledge about prehistory. Archaeology consists of the use of surveying and excavation, as well as further analysis of statistics, to discover more about our past [1].

In a variety of fields, the term 'nondestructive' has different definitions. However, all fields agree that information is scientifically important [2]. Archeology has traditionally centered around external physical characteristics as well as the inner chemical systems of samples; however, nondestructive evaluation on an objective (as opposed to theoretical) basis is strongly encouraged by the international businesses and academic institutions entrusted with curatorial responsibility for cultural history [3].

3.1 Introduction

Survey strategies include nondestructive testing, which is the best way to obtain more information about a given cultural heritage. This can be done without destroying it through the use of techniques that have been developed over time. Major factors in archaeological research and restoration are the abilities to gauge the amount of deposits and to assess the state of an artifact without disrupting it [4].

In order to facilitate analysis of the cultural background of objects, some analytical techniques permit observation without requiring the removal of a sample from an item. These techniques may be classified depending on their level of invasiveness ranging from noninvasive to damaging. Some of the techniques used

doi:10.1088/978-0-7503-4846-1ch3

most often are noninvasive spectroscopies, which include ultraviolet–visible (UV–Vis) spectroscopy [5], infrared spectroscopy [6], Raman spectroscopy [7], and x-ray fluorescence [8]. These devices provide complementary data that helps us to deduce an object's chemical identity and morphology, though many strategies can be applied over the course of a technical observation to help in understanding an object's constituent substances. X-ray fluorescence provides elemental data, while infrared spectroscopy offers information about chemical bonding, which can be used for molecular identification [9].

In the previous edition of this book [10], we reviewed the concept of non-destructive analysis of archaeological heritage samples and the types and forms of device used for this purpose. In this chapter, we review some models and special cases in which devices for the nondestructive analysis of samples have been used; this provides a further demonstration of the importance of the non-destructive testing technique in the archaeological world. Briefly, we can say that each archaeological specialist pursues the maximization of information with the minimization of consumed samples.

One important component in studying archaeological materials is identifying the type of an artifact. This first step helps to identify how it was made, its function, when it might have been made, and more information about the culture that produced it. Archaeometric research involves the integration of physics, chemistry, and materials science. It is important to understand that every method has its benefits and flaws. A broad perspective should be considered in order to understand the information gained through research [11].

3.2 Mobile analytical solutions for art conservation and archaeometry

It can be difficult to transport artifacts because of practical and political constraints. Thus, portable instrumentation becomes a necessity because it can be used to study many different types of sample. For example: you need to examine the focus of your project, but it cannot be removed from a high-profile gallery; it may require scaffolding to reach certain areas; or the location of the testing may be outdoors at the next excavation site. When you have to take the laboratory to the target of your project, cellular and portable instrumentation needs to be available[1]. In recent years, different instruments have been used in the study of archaeological sites to get better insights into what is happening within their environments. For example, the metals that are found at a particular site need to be kept within tightly regulated conditions, i.e. held at constant temperature and humidity, because they are so sensitive to change. It can be disruptive to move these objects to a laboratory where their environmental stability will be disrupted [12, 13]. To address this situation, researchers have developed equipment that can be transported and used *in situ*, so that the studied objects are not burdened by any physical, chemical, or mechanical

[1] https://www.bruker.com/en/applications/academia-materials-science/art-conservation-archaeology/the-mobile-lab.html

stresses. Such equipment is necessary not only for the physics of art and archaeology but for other disciplines as well.

3.2.1 Portable x-ray fluorescence

Portable x-ray fluorescence (PXRF) units have been extremely beneficial in the growth of packages for archaeology and history of art, as they assist in avoiding destruction and complying with archaeologists' call for noninvasive sampling. Many worldwide applications of PXRF to cultural materials consisting of ceramics, clays, soils, painted surfaces, and metals have demonstrated its value and have produced useful results. PXRF creates a unique opportunity for the non-detrimental study of historic substances, and portability is vital in making this technique powerful. Big steps forward have been taken in this field: spectrometers weighing barely a few kilograms have been built, benefiting from the use of miniaturized x-ray tubes and thermo-electrically cooled detectors. This advancement in portable devices competes with, and in many cases, acts independently of cumbersome laboratory contraptions [14].

Handheld x-ray fluorescence spectroscopy has supported great advances in the study of cultural storage materials, including speed and noninvasiveness when examining works of art, artifacts in their original sites, and in museums or conservation studios. X-ray fluorescence (XRF) spectroscopy is a technique used to determine the fundamental compositions of a huge variety of substances. In recent years, developments in x-ray emission and detection have led to the production and commercialization of handheld and other types of portable XRF device, which can conduct speedy and reliable analysis without having to take away sensitive and precious items from a museum/conservation studio. Analyzing artistic endeavors or works of art and collections *in situ* eliminates the need for fragile objects to be moved from a museum/conservation studio to the laboratory [15].

Over the past 30 years, PXRF analyzers have been ideal equipment for use in the growing cultural heritage applications used for research in archaeology and geo-archaeology, including:

(i) on-site material characterization using handheld systems
(ii) museum evaluation via handheld and computer systems (PXRF can identify additives in the pigments used in paintings and glazes and steel alloy constituent materials; it can characterize items such as jewelry, silverware, and weaponry)
(iii) in the field of conservation, technological expertise with the objective of matching pigments in a speedy and accurate way, supporting awareness of how items have been preserved in the past

XRF has been utilized in combination with Raman spectroscopy for the identification of mineral pigments in cutting-edge artwork [16, 17] and binding media and varnishes utilized in medieval paintings and manuscripts [16]. Moreover, the combination between XRF and Raman spectroscopy could be used in detecting fake forged artifacts. The use of PXRF evaluation on a variety of artifacts based totally on elemental data has been extraordinarily beneficial [18, 19]. PXRF makes it possible to guarantee authenticity by capturing information such as provenance and the materials used to create an artifact, which helps to identify specific pieces.

Handheld XRF has been used to analyze the chemical composition of a sunken ship's book cover (a warship which sank in March 1665) and to identify gold and red pigments [20]. The evaluation showed the presence of gold (Au) and detected lead (Pb) and mercury (Hg) within the purple paint, suggesting the usage of cinnabar/vermilion (herbal and artificial HgS, respectively) and crimson lead (Pb_3O_4) pigments. Furthermore, PXRF was used to research some intact metallic artifacts recovered from the excavation of a Dutch East India Company ship which was wrecked off the Kent coast in 1740 to identify the production method, the alloy composition, and the country of origin, in an effort to determine the most suitable conservation techniques [21].

Vincenza Crupi and Sebastiano D'Amico *et al* have proved the effectiveness of multiple-technique *in situ* investigations, which enable one to take advantage of complementary facts based on an exhaustive characterization of artifacts without the need to sample or destroy them. The researchers' choice of site was Scifi in Italy, where they analyzed pottery fragments and dated them back to the 3rd century BCE and 4th–5th century CE. They used XRF, Raman, and infrared testing strategies, which characterized pre-existing pottery from an archaeological perspective for the first time for cataloging purposes [22].

3.2.2 Portable Fourier-transform infrared spectroscopy–attenuated total reflectance

Fourier-transform infrared spectroscopy (FTIR) has been utilized on bones to obtain information about both the organic and inorganic components of a sample. It is an important tool for archaeologists because it allows them to understand the materials that have altered the bone diagenesis [23, 24]. Studies have used portable FTIR devices to identify the minerals in archaeological sediments, organic compounds, and bone diagenesis [25, 26]. The main benefit of a transportable tool, as mentioned by Weiner *et al* [27], is that it allows editing, sampling, and excavation strategies to be used during an ongoing task; for this reason, it saves time and reduces the number of sampling attempts, it allows the queries pursued by the study to be adjusted, and it allows archaeological problems to be resolved in keeping with the taphonomic context.

Geneviève Pothier *et al* [28] produced a research paper that presented evidence to support the concept of comparing collagen maintenance in a quick and minimally damaging manner using this methodology prior to in-depth evaluation by the use of a portable FTIR device equipped with an attenuated total reflectance (ATR) accessory. This approach was tested on faunal assemblages from north-western Italian sites: Riparo Bombrini and Arma Veirana. The results indicated that FTIR devices are amazing tools with which to investigate collagen preservation in fragmented archaeological bones.

3.2.3 Raman spectroscopy

Raman spectroscopy has various benefits; for example, it is non-detrimental, sensitive, and portable. As a result, it has been the most popular analysis tool for cultural items in recent years [29, 30]. Depending on the requirement, Raman spectroscopy can be replaced or augmented by additional molecular instrumentation, such as FTIR and fiber-optic reflectance spectroscopy (FORS) [31, 32]. Even if *in situ* analysis provides the essential information required to identify the degradation processes affecting items, these results sometimes need to be supported by laboratory analysis in order to understand, for example, their stratigraphic and surface distribution [33].

The use of Raman spectroscopy for archaeological analysis has focused on paints and pigments, resins and pitch, and plaster-like substances [34, 35]. It can be used to detect each organic and inorganic material and has gained a reputation because of its nondestructive nature. However, background fluorescence might limit the sensitivity of this approach, and archaeological substances may additionally be subject to taphonomic processes that make it difficult to fit the results to those of popular reference samples [36]. Moreover, until lately, Raman evaluation has been laboratory orientated.

In the field research into works of art, there has long been a requirement for on-site analysis, especially for treasured artworks held under strict museum security and for wall artworks which cannot be bodily removed from their settings; further, the use of portable Raman spectroscopy in archaeological and geological paintings as a first-step screening device which obviates the need for further screening and wasteful specimen collection is high on the wish lists of student spectroscopists [37].

The desire to perform *in situ* studies is exemplified by the management of wall artworks: while sampling is not always allowed, the simplest way to observe such artifacts is to bring the instruments to the wall artwork. However, when paintings are hard to access, for example, if they are located on the ceiling of a chapel or a cathedral, the system has to be carried to the top of the scaffolding. In such cases, obtaining the best possible spectra is a difficult task; scaffolding is commonly not completely stable, and if one of the operators moves, this influences the precise positioning (and focus) of the probe head. In the study of wall paintings, Raman spectroscopy is often combined with other analytical methods, such as XRF analysis. Examples of this approach include:

- the analysis of mural artworks in the Aosta valley (Italy) [38],
- the analysis of vault paintings at Antwerp Cathedral (Belgium) [39],
- the evaluation of ancient Egyptian wall paintings in a tomb near Luxor [40], and
- the investigation of wall paintings in the Democratic People's Republic of Korea performed by Mazzeo *et al* [41] using Raman spectroscopy.

In addition to identifying the pigments in a painting, Pérez-Alonso *et al* [42] identified oxalates via *in situ* Raman spectroscopy, and it has now been established that it can be used to investigate the degradation mechanisms of an artwork. Several studies have also investigated antique mortars using Raman spectroscopy, which captures various microscopic details with great sensitivity, such as biodegradation and other structural changes occurring on the microscale; for example, Boschetti *et al* [43] studied the composition of the mortars used in ancient Roman mosaics.

Raman spectroscopy is often used to describe the molecular structures of glasses. Using a Raman spectroscope and different types of laser, it is possible to measure the molecular structures of blue and green glasses. Researchers have determined that a Raman spectroscope can also be used to measure the shape of silicate matrices [44].

Raman spectroscopy filters out both organic and inorganic substances; thus, new synthetic pigments can be easily distinguished from traditional pigments made primarily from ground minerals. Raman spectroscopy has been applied to a variety of matrices and is suited to the study of complex assemblages in artworks [45, 46], ceramics [47, 48], manuscripts [49, 50], gemstones [51], glass [52], wall paintings [53, 54], and rock art [55]. These studies provide records that may be used to date artworks and to identify counterfeits. Raman spectroscopy has also used to investigate the impacts of environmental factors and restorations.

In addition, Maitane Olivares *et al* [56] conducted a study in which special flint samples from numerous sites of geological and archaeological importance in the Basque Country were collected and analyzed. The use of Raman spectroscopy made it possible to distinguish between quartz and mullite in chert samples containing artificial materials, and showed that the ratio of these two components was related to the availability of chert. Therefore, they concluded from their study that the combination of vibrational spectroscopy strategies with other fundamental strategies, such as energy-dispersive x-ray fluorescence (EDXRF), may be a powerful strategy when the characterization of chert or similar geological materials is required. Most of these methods allow for a benign chemical assessment, enabling the characterization of precious archaeological samples.

3.3 Activation analysis techniques

Activation analysis is a notably sensitive nondestructive approach for the qualitative and quantitative determination of the atomic composition of a sample. It has been particularly beneficial for the determination of the components of complex samples (minerals, environmental samples, organic and archaeological items, and so on), as it offers an easy alternative to much more difficult, tedious, and damaging strategies. Its greatest challenge is the requirement for a strong irradiation source [57].

In various fields in which there is a requirement not to damage the tested material, particularly within the fields of archeology and forensics, activation analysis is considered to be nondestructive [58].

Activation analysis consists of a set of strategies that differ depending on which particle is used to cause reactions in the material under test. These strategies are just like those used in XRF and PIXE, except that they use higher-power incident particles. These techniques use charged particles (including protons and helium) in charged particle activation analysis (CPAA), neutrons in neutron activation evaluation (NAA) [59, 60], and photons in photon activation analysis (PAA) [61].

3.3.1 Photon activation analysis and neutron activation evaluation

PAA is just like NAA: at its simplest, it needs a few common systems consisting of detectors and a linear accelerator. It can be a useful replacement; the recent work by Borgwardt [62] *et al* have demonstrated a few of its capabilities. This study found 31 elements, with tellurium detected at 150 ng/g. PAA's spectrum and detection limits, as well as its bulk analysis capabilities, make it ideal for paleontology studies. As opposed to ionizing the atom, these techniques stimulate the nucleus of the atom, causing protons and neutrons (or combinations of these) to be ejected from the nucleus. This transmutes the atom into a special, often radioactive element, which eventually decays, giving off active photons that can be detected. This transmutation, in addition to the triggered radioactivity, offers a completely unique outlook on nondestructive analysis. NAA is additionally considered to be particularly nondestructive in forensics, as it requires no 'blank' [63]. PAA offers this benefit, in addition to the common advantages with NAA. It can mechanically analyze 30+ components in a sample with ordinary detection limits in the ppm range. In a few cases, detection limits at the ppb level or even less are possible [64]. The accuracy of the technique has been proven to be comparable to, or higher than that of NAA [65]. Similarly to those, PAA offers the unique advantage of performing complete pattern evaluation. The penetration depth of photons permits huge samples to be measured without problems. The entire composition of inhomogeneous samples can be measured effortlessly. This is particularly useful for fields such as archeology, wherein the surfaces of samples generally tend to be subjected to various processes that affect their composition [66]. To avoid this, other techniques require sampling of the item or abrasion of the surface, which are undesirable. Even though one may accept such an undesirable analysis, the pattern can be inhomogeneous, hence the analysis might be suboptimal.

In comparison to NAA, PAA offers several benefits:

First: is the equipment required; NAA calls for a nuclear reactor, the provision of which is predicted to decrease in the future [60], whilst PAA requires a particle accelerator [59, 67].

Second: PAA makes use of high-power photons as an irradiation supply. This offers the technique large penetration as compared to other techniques, allowing the bulk analysis of large samples, i.e. hundreds of grams or more, without other special requirements.

Furthermore: the capacity to manipulate the radiation supply allows many large samples to be analyzed using devices such as collimators and electron beam scanners [68].

In a study, Tyler C Borgwardt *et al* [62] observed that PAA has been proven to be a capable nondestructive method for paleontological research. As a tool for analyzing the elemental composition of fossils and rocks, it is a broad-spectrum multiple-detail analysis device. This work recognized 31 elements, the smallest amount of which was detected at a level of 150 ng g^{-1} for tellurium. PAA's multielement spectrum and exceptional detection limits, together with its capability to perform bulk analysis of large samples, make it suitable for various kinds of paleontological study. The nondestructive approach of PAA makes it a smart alternative to other less favorable contemporary evaluation strategies.

3.3.2 Prompt gamma activation analysis

Prompt gamma activation analysis (PGAA) is a nondestructive bulk analytical technique primarily based at the detection of characteristic gamma photons emitted in (n, γ)

reactions [69] when a sample is irradiated with a beam of cold neutrons produced by a research reactor. Considering that neutrons penetrate to a depth of many centimeters in materials, PGAA offers bulk compositional analysis for the entire irradiated volume, which can include the entire depth of a small object. PGAA is able to detect all of the chemical elements except helium. In most cases, primary additives and some of the commoner trace elements may be quantified [70].

PGAA is based on the radioactive capture of neutrons by cores. It is a unique test for the nondestructive detection of hydrogen. Its main points of interest are its nondestructive nature and the absence of test requirements [71]. Zs. Kasztovszky *et al* attempted to examine silver–copper coins originating from Hungary in the 16th and 20th centuries as well as Roman bronze fibulas from the first and fourth centuries AD using PGAA. They demonstrated that PGAA is suitable for the quantitative assessment of diverse archaeological steel tests. The nondestructive and multielement capabilities of PGAA offer key benefits for archaeological investigations.

Kluge and coworkers used PGAA to study two ring-like Celtic burial gifts from Bavaria. The results clearly confirmed completely unique compositions and internal object structures [721]. The chemical constitutions of 48 glass items excavated at Histria and Tomis, Romania, mainly dated to the first–fourth centuries AD, were determined by the use of triggered gamma activation analysis at the Budapest Neutron Center (BNC) by Roxana Bugoi *et al* [73]. The use of the gamma activation analysis information provided clues pointing to glass recycling, which were evidenced by numerous samples. Archaeological and compositional arguments support the idea that during the whole of the early Roman Empire, recycling practices were used in the secondary glass workshops at Tomis.

3.4 Electron paramagnetic resonance

Electron paramagnetic resonance (EPR), also referred to as electron spin resonance (ESR), is increasingly being utilized in archaeology, and the sorts and functions of the materials analyzed increase every year; dating is one of its most vital applications [74].

Charles Dutoit, Didier Gourier *et al* [75] sought to use the noninvasive EPR method to discover two additives of bitumen (vanadyl porphyrins and carbonaceous radicals) which used to be imbedded in the embalming fabric of ancient Egyptian sarcophagi (or coffins). This could provide data about the presence, origin, and processing of bitumen inside the embalming fabric used by the ancient Egyptians. The researchers gathered samples of black matter from different ancient Egyptian coffins: two human mummies and four animal mummies dated from 744 to 30 BC. They investigated these samples by means of EPR and then compared the results thus obtained to those of reference bitumen samples. The research group observed that the relative amounts of vanadyl compounds and carbonaceous radicals ought to allow them to distinguish between bitumens of marine origin (Dead Sea) and land-plant origin (tar pit). In addition, they revealed vanadyl compounds which mostly formed as a reaction product of the vanadyl porphyrins and other embalming additives[2].

3.5 Ground-penetrating radar

Ground-penetrating radar (GPR) could be a noninvasive geophysical strategy for high-resolution imaging and the characterization of subsurface media through the transmission and reception of the high-frequency band (approximately between 10 MHz and 2600 MHz) [76–78] of electromagnetic waves. This approach is likewise used for the nondestructive testing of bulk stone in quarries as described by Lualdi and Zanzi [79] and in pre-excavation archaeological investigations as reported by Conyers and Leckebusch [80]. Within the field of social built legacy, GPR has progressively been utilized for the appraisal of the conservation state of craftsmanship and monuments [81–83]. Furthermore, this approach has the ability to locate small items at incredible depth and to distinguish between the intensities of the signal origins. GPR is used to discover and map subsurface archaeological artifacts, structures, and patterning [84].

Nondestructive geophysical methods are regularly utilized in combination with each other [85]. Therefore, for example, the search for archaeological deposits is conducted in its early stages through the use of GPR and electrical techniques. In fact, in the interpretation phase, the ambiguities contained within the results obtained from a single approach can frequently be eliminated by considering the results acquired using a second method. As an example, the reflections in a GPR survey due to the presence of a wall and/or a tomb are comparable. By integrating the GPR survey with an electrical survey, this ambiguity can be resolved by considering the surprisingly large resistivity values that might be associated with the wall in parallel with the especially low resistivity associated with the earth-filled tomb [86].

The checking of old heritage objects can be an task of great interest and the highest significance for their practicable maintenance and restoration. In particular, an understanding of their internal shapes is the key to their preservation and/or rebuilding. In the case of buildings, such shapes are composed of the distinctive shapes of stones or bricks with timber, iron, or many other components (such as ties) embedded into walls and cavities [87]. Moreover, the wall thickness and the frequency of separation and cracking is important in confirming the stability of homes. In the case of examinations of cultural built history, the foremost problem is associated with the impossibility of touching and taking samples that can be studied within the laboratory or maybe *in situ*. The assessment of the degree of decay degree can be performed using various methods and with different levels of accuracy. In general, it can be carried out by qualitative and quantitative methods.

Aleksandar ristíc *et al* made effective use of GPR to reveal the remains of the Württemberg-Stambol gate in the subsurface of Republic Square in Belgrade, Serbia [88].

[2] https://www.heritagedaily.com/2020/12/a-non-destructive-method-for-analyzing-ancient-egyptian-embalming-materials/136502

Based on the results of the GPR study, their advice is to incorporate GPR as a habitual discipline in construction tasks within historic towns. GPR maps and images can offer very beneficial information and guide the positions of excavations or outline sensitive regions containing archaeological sites to be avoided in order to maintain and defend our cultural historical heritage. In the case of excavation works, GPR images also make it possible to estimate the amount of work involved; thus a GPR survey may be be wanted both to complete the excavation works and to develop a preliminary concept of the reconstruction possibilities.

A GPR system essentially measures strength of the signal reflected or scattered by an object; the amplitudes are recorded as a function of journey time. The system contains a signal generator, transmitting and receiving antennae, and a receiver that may or may not have a recording facility or the ability to output graphical hard copy. A few superior systems have an onboard computer that helps to process information at the same time as it acquires records inside the area and publishes the recording. There are three modes of radar deployment [89]:

(i) reflection profiling using the commonplace offset mode,
(ii) Wide-angle reflection and refraction (WARR) or common-midpoint (CMP) sounding,
(iii) trans illumination or radar tomography.

The depth of penetration of a radar pulse depends to a large degree on the electrical conductivity of the tested artifact and the frequency of the antenna used. In general, the amount of penetration decreases with increasing conductivity. Furthermore, for a given object, a decrease in antenna frequency increases penetration depth; however, the resolution is thereby reduced. Therefore, a compromise ought to be reached between the penetration depth and the resolution.

3.6 Future scope

At the moment, the term 'nondestructive' analysis is used very loosely and subjectively. Many techniques are generally nondestructive but best used under ideal conditions as defined by sample size, detection limits, and many others. More essential evaluations need to be finished earlier rather than claiming that a method is nondestructive for specific research. Although no method is clearly nondestructive in all situations, a crucial analysis can assist in generating area-specific, or minimally unfavorable, definitions of use. It is necessary to develop and organize nondestructive strategies and measurement equipment which can routinely be used with historic artifacts of all kinds.

In this chapter, I tried to shed light on the nondestructive techniques most commonly used in archaeometric studies and to describe some case studies for each technique to provide further clarification.

References

[1] Khelifi A 2018 The archeology field in the mobile era: a roadmap to catch-up *Int. J. Inf. Manage.* **12** 73–94

[2] Riquelme F, Ruvalcaba Sil J L and Alvarado-Ortega J 2009 Palaeometry: non-destructive analysis of fossil materials *Bol. Soc. Geol. Mex.* **61** 177–83
[3] European Commission, Directorate-General for Research and Innovation 2000 *Ion beam study of art and archaeological objects : A contribution by members of the COST G1 Action* EUR 19218 Office for Official Publications of the European Commission Eds: G Demortier, A Adriaens https://op.europa.eu/s/ywUI
[4] Biró K T 2005 Non-destructive research in archaeology *J. Radioanal. Nucl. Chem.* **265** 235–40
[5] Gabrieli F, Dooley K A, Facini M and Delaney J K 2019 Near-UV to mid-IR reflectance imaging spectroscopy of paintings on the macroscale *Sci. Adv.* **5** eaaw7794
[6] Rosi F, Cartechini L, Sali D and Miliani C 2019 Recent trends in the application of Fourier transform infrared (FT-IR) spectroscopy in heritage science: from micro- to non-invasive FT-IR *Phys. Sci. Rev.* **4** 20180006
[7] Brosseau C L, Rayner K S, Casadio F, Grzywacz C M and Van Duyne R P 2009 Surface-enhanced Raman spectroscopy: a direct method to identify colorants in various artist media *Anal. Chem.* **81** 7443–7
[8] Dooley K A, Conover D M, Glinsman L D and Delaney J K 2014 Complementary standoff chemical imaging to map and identify artist materials in an early Italian Renaissance panel painting *Angew. Chem.* **53** 13775–9
[9] Olivia R K and Joseph P H 2022 An introduction to the electron paramagnetic resonance spectral library of pigments *Heritage* **5** 545–66
[10] Osman E 2020 Spectrometry as a non-destructive technique in identifying cultural archaeological heritage *Spectroscopic Techniques for Archaeological and Cultural Heritage ResearchIOP Series in Spectroscopic Methods and Applications* ed A Kumar (Bristol: IOP Publishing) ch 1 pp 1–12
[11] Laura T, Ayed B A, Nadia C, Rémy C, Cătălin D, Sorin C, Claudiu T, Carol T and Mărioara A 2021 Characterization of archaeological artefacts using methods specific to materials science: the case study of dacian ceramics from 2nd C. BC to 1st C. AD *Materials* **14** 1–11
[12] Dillmann P, Mazaudier F and Hœrlé S 2004 Advances in understanding atmospheric corrosion of iron. I. Rust characterization of ancient ferrous artefacts exposed to indoor atmospheric corrosion *Corros. Sci.* **46** 1401–29
[13] Tamura H 2008 The role of rusts in corrosion and corrosion protection of iron and steel *Corros. Sci.* **50** 1872–83
[14] Liritzis I and Zacharias N 2010 X-ray fluorescence spectrometry (XRF) in geoarchaeology *Portable XRF of Archaeological Artifacts: Current Research, Potentials and Limitations* ed M S Shackley (New York: Springer) p 6
[15] Gherardi F 2021 Hand-held X-ray fluorescence analysis of archaeological artefacts: challenges, advantages and limitations *Anal. Methods* **13** 3731
[16] Vandenabeele P, Wehling B, Moens L, Edwards H, De Reu M and Van Hooydonk G 2000 Analysis with micro Raman spectroscopy of natural organic binding media and varnishes used in art *Anal. Chim. Acta* **407** 261–74
[17] Liritzis I and Polychroniadou E 2007 Optical and analytical techniques applied to the Amfissa Cathedral mural paintings made by the Greek artist Spyros Papaloukas (1892–1957) *Rev. Archaeometrie (Archaeosciences)* **31** 97–112

[18] Appoloni C R, Blonski M S, Parreira P S and Souza L A C 2007 Pigments elementary chemical composition study of a gainsborough attributed painting employing a portable X-rays fluorescence system *AIP Conf. Proc.* **884** 459–64
[19] Guerra M F 2008 An overview on the ancient Goldsmith's skill and the circulation of gold in the past: the role of X-ray based techniques *X-Ray Spectrom.* **37** 317–27
[20] July 2020 70–81 The London wreck—a kaleidoscope of specialists, materials and artefacts. Historic England Research https://historicengland.org.uk/whats-new/research/back-issues/the-london-wreck-a-kaleidoscope-of-specialists-materials-and-artefacts/
[21] Kuiper E 2018 Conservation of copper alloy artefacts from the Dutch East Indiaman the Rooswijk *Research Department Reports* (Historic England) 36 https://research.historicengland.org.uk/
[22] Vincenza C *et al* 2018 Mobile spectroscopy in archaeometry: some case study *J. Spectrosc.* **2018** 8295291
[23] Chadefaux C, Le Hô A S, Bellot-Gurlet L and Reiche I 2009 Curve-fitting micro-ATRFTIR studies of the amide I and II bands of type I collagen in archaeological bone materials *e-Preserv. Sci.* **6** 129–37 http://www.morana-rtd.com/e-preservationscience/2009/Chadefaux-04-07-2008.pdf
[24] Thompson T J U, Gauthier M and Islam M 2009 The application of a new method of Fourier transform infrared spectroscopy to the analysis of burned bone *J. Archaeol. Sci.* **36** 910–4
[25] Toffolo M B, Brink J S and Berna F 2015 Bone diagenesis at the Florisbad Spring Site, Free State Province (South Africa): implications for the taphonomy of the middle and late Pleistocene faunal assemblages *J. Archaeol. Sci. Rep.* **4** 152–63
[26] Monnier G F 2018 A review of infrared spectroscopy in microarchaeology: methods, applications, and recent trends *J. Archaeol. Sci. Rep.* **18** 806–23
[27] Weiner S 2010 *Microarchaeology: Beyond the Visible Archaeological Record* (Cambridge: Cambridge University Press)
[28] Geneviève P B, Susan M M, Julien R S, Jamie H, Christopher E M, Fabio N, Roy W and Michael B 2019 Portable FTIR for on-site screening of archaeological bone intended for ZooMS collagen fingerprint analysis *J. Archaeolog. Sci.* **26** 101862
[29] Colomban P, Tournié A, Maucuer M and Meynard P 2012 On-site Raman and XRF analysis of Japanese/Chinese bronze/brass patina—the search for specific Raman signatures *J. Raman Spectrosc.* **43** 799–808
[30] Gómez-Laserna O, Olazabal M A, Morillas H, Prieto-Taboada N, Martinez-Arkarazo I, Arana G and Madariaga J M 2013 In situ spectroscopic assessment of the conservation state of building materials from a palace house affected by infiltration water *J. Raman Spectrosc.* **44** 1277–84
[31] Ploeger R, Chiantore O, Scalarone D and Poli T 2011 Mid-infrared fiber-optic reflection spectroscopy (FORS) analysis of artists' alkyd paints on different supports *Appl. Spectrosc.* **65** 429–35
[32] Cheilakou E, Troullinos M and Koui M 2014 Identification of pigments on Byzantine wall paintings from Crete (14th century AD) using non-invasive fiber optics diffuse reflectance spectroscopy (FORS) *J. Archaeol. Sci.* **41** 541–55
[33] Marco V, Ilaria C, Silvia Fdez-Ortiz V, Laura G, Iñaki G, Kepa C, Agustín A and Juan M M 2016 Study of corrosion in archaeological gilded irons by Raman imaging and a coupled scanning electron microscope–Raman system *Phil. Trans. R. Soc.* A **374** 20160046

[34] Edwards H G M and Chalmers J M 2005 *Raman Spectroscopy in Archaeology And Art History* (London: Royal Society of Chemistry)
[35] Malainey M E 2011 *A Consumer's Guide to Archaeological Science: Analytical Techniques* (Berlin: Springer)
[36] Smith G D and Clark R J H 2004 Review article: Raman microscopy in archaeological science *J. Archaeolog. Sci.* **31** 1137–60
[37] Vandenabeele P, Edwards H G M and Jehlička J 2014 Review article: the role of mobile instrumentation in novel applications of Raman spectroscopy: archaeometry, geosciences, and forensics *Chem. Soc. Rev.* **43** 2628–2649
[38] Lorenzo A, Davide V, Valentina C, Maurizio A and Piero M 2009 Combined use of FORS, XRF and Raman spectroscopy in the study of mural paintings in the Aosta Valley (Italy) *Anal. Bioanal. Chem.* **395** 2005–13
[39] Annelien D, WalterSchudel M V, Helena W, Anna B, Peter V and Luc M 2010 In situ investigations of vault paintings in the Antwerp Cathedral *Spectrochim. Acta* A **75** 511–9
[40] Hussein M 2013 Raman microscopic analysis of a multi-pigmented surface from the Theban Tomb (TT277), Luxor, Egypt *Acta Phys. Pol.* A **123** 4
[41] Peter V, Karel L, Sofie M, Walter S, Anna B and Luc M 2005 In situ analysis of mediaeval wall paintings: a challenge for mobile Raman spectroscopy *Anal. Bioanal. Chem.* **383** 707–12
[42] Maite P A, Kepa C and Juan M M 2006 Investigation of degradation mechanisms by portable Raman spectroscopy and thermodynamic speciation: the wall painting of Santa María de Lemoniz (Basque Country, North of Spain) *Anal. Chim. Acta* **571** 121–8
[43] Cristina B, Anna C and Pietro B 2008 Raman characterization of painted mortar in Republican Roman Mosaics *J. Raman Spectrosc.* **39** 1085–90
[44] Philippe C and Ove P 2005 Non-destructive determination of the structure and composition of glazes by Raman spectroscopy *J. Am. Ceram. Soc.* **88** 390–5
[45] Gutiérrez-Neira P C, Agulló-Rueda F, Climent-Font A and Garrido C 2013 Raman spectroscopy analysis of pigments on Diego Veláquez paintings *Vib. Spectrosc.* **69** 13–20
[46] Proniewicz E E, Szmelter-Fausek B, Olszewska-Świetlik J and Proniewicz L M 2014 Micro-Raman spectroscopy analysis of the 17th century panel painting 'Servilius Appius' by Isaac van den Blocke *J. Raman Spectros* **1019** 11–2
[47] Striova Lofrumento J C, Zoppi A and Castellucci E M 2006 Prehistoric Anasazi ceramics studied by micro-Raman spectroscopy *J. Raman Spectrosc.* **37** 1139
[48] Colomban P 2004 Raman spectrometry, a uunique tool to analyze and classify ancient ceramics and glasses *Appl. Phys.* **79** 167
[49] Muralha V S F, Burgio L and Clark R J H 2012 Raman spectroscopy analysis of pigments on 16–17th c. Persian manuscripts *Spectrochim. Acta* **92** 21
[50] Deneckere A, Leeflang M, Bloem M, Chavanne-Mazel C A, Vekemans B, Vincze L, Vandenabeele P and Moens L 2011 The use of mobile Raman spectroscopy to compart three full-page miniatures from the breviary of Arnold of Egmond *Spectrochim. Acta* A **83** 194
[51] Reiche I, Pages-Camagna S and Lambacher L 2004 In situ spectroscopy investigations of the adorning gemstones on the reliquary *Heinrich's cross* from the treasury of Basel Cathedral *J. Raman Spectrosc* **719** 8–9
[52] Caggiani M C, Colomban P, Valotteau C, Mangone A and Cambon P 2013 Mobile Raman spectroscopy analysis of ancient enameled glass masterpieces *Anal. Methods* **5** 4345

[53] Zucchuatti A, Prati P, Bouquillon A, Giutini L, Massi M, Migliori A, Cagnana A and Roascio S 2004 Characterization of early medieval frescoes by µ-PIXE, SEM and Raman spectroscopy *Nucl. Instrum. Methods Phys. Res.* B **219–220** 20–5

[54] Edwards H G M, Farwell D W and Rozenberg S 1999 Raman spectroscopy study of red pigment and fresco fragments from King Herod's Palace at Jericho *J. Raman. Spectrosc.* **30** 361

[55] Tournié A, Prinsloo L C, Paris C, Colomban P and Smith B 2011 The first *in situ* Raman spectroscopic study of San rock art in South Africa: procedures and preliminary results *J. Raman Spectrosc* **42** 399

[56] Olivares M, Tarrino A, Murelaga X, Aceta J I, Castroa K and Etxebarria N 2009 Non-destructive spectrometry methods to study the distribution of archaeological and geological chert samples *Spectrochim. Acta* A **73** 492–7

[57] Choppin G R and Jan-Olov L J 2002 Uses of radioactive tracers *Radiochemistry and Nuclear Chemistry* 3 (Waltham, MA: Butterworth-Heinemann) p 9

[58] Borgwardt T C and Wells D P 2017 What does non-destructive analysis mean *Cogent Chem.* **3** 1–6

[59] Glascock M D and Neff H 2003 Neutron activation analysis and provenance research in archaeology *Meas. Sci. Technol.* **14** 1516–26

[60] Tykot R H 2002 Chemical fingerprinting and source tracing of obsidian: the central mediterranean trade in black gold *Acc. Chem. Res.* **35** 618–27

[61] Segebade C, Starovoitova V N, Borgwardt T and Wells D 2017 Principles, methodologies, and applications of photon activation analysis: a review *J. Radioanal. Nucl. Chem.* **312** 443–59

[62] Borgwardt T C and Wells D P 2017 What does non-destructive analysis mean? *Cogent Chem.* **3** 1405767

[63] Sudersanana M, Kayasth S R, Pant D R, Chattopadhyay N and Bhattacharyya C N 2002 Application of nuclear and allied techniques for the characterization of forensic samples *Advances in Destructive and Non-Destructive Analysis for Environmental Monitoring and Nuclear Forensics* (Karlsruhe: IAEA), 120–8 https://www-pub.iaea.org/mtcd/publications/pdf/pub1169_web.pdf

[64] Schulze D and Segebade C 2005 Activation analysis—photon activation *Encyclopedia of Analytical Science* 2nd edn ed P Worsfold, A Townshend and C Poole (Oxford: Elsevier) pp 20–7

[65] Řanda Z, Kučera J, Mizera J and Frána J 2007 Comparison of the role of photon and neutron activation analyses for elemental characterization of geological, biological and environmental materials *J. Radioanal. Nucl. Chem.* **271** 589–96

[66] Caneva C and Ferretti M 2000 In XRF spectrometers for non-destructive investigations in art and archaeology: the cost of portability *Proc. 15th World Conf. on Non-Destructive Testing* (Brescia: Italian Society for Nondestructive Testing)

[67] DOE 2010 *Accelerators for America's Future* (New York: U.S. Department of Energy) https://science.osti.gov/-/media/hep/pdf/files/pdfs/Accel_for_Americas_Future_final_report.pdf

[68] Borgwardt T C 2018 Non-destructive photon activation analysis in paleontology *Proc. IECG 2018* **2** 564

[69] Révay Z, Firestone R B, Belgya T and Molnár G L 2004 Prompt gamma-ray spectrum catalog , *Handbook of Prompt Gamma Activation Analysis with Neutron Beams* ed G L Molnár (Dordrecht: Kluwer Academic) pp 173–364

[70] Atkinson D and Rimmer M 2014 The use of neutron analysis technique for detecting the concentration and distribution of chloride ions in archaeological iron *Archaeometry* **56** 841–59

[71] Kasztovszky Z, Revay Z, Belgya T and Molnar G L 2000 Nondestructive analysis of meatls by PGAA at the Budapest Research Reactor *J. Radioanl. Nucl. Chem.* **244** 379–82

[72] Kluge E J, Stieghorst C, Wagner F E, Gebhard R, Révay Z and Jolie J 2018 Archaeometry at the PGAA facility of MLZ—prompt gamma-ray neutron activation analysis and neutron tomography , *J. Archaeol. Sci. Rep.* **20** 303–6

[73] Roxana B, Alexandra T, Veronika S, Ildikó H, Lauren¸ Tiu C, Irina A and Zsolt K 2022 Shedding light on Roman glass consumption on the western coast of the Black Sea *Materials* **15** 403

[74] Laurent L P and Graham A W 2016 Application of EPR in studies of archaeological samples *Modern Magnetic Resonance* 2nd edn (Cham: Springer) pp 1–25

[75] Charles E D, Laurent B, Hitomi F, Agnes L D and Didier G 2020 Nondestructive analysis of mummification balms in ancient Egypt based on EPR of vanadyl and organic radical markers of bitumen *Anal. Chem.* **92** 15445–53

[76] Martinho E and Dionísio A 2014 Main geophysical techniques used for non-destructive evaluation in cultural built heritage: a review *J. Geophys. Eng.* **11** 053001

[77] Liu Y, Shi Z, Wang B and Yu T 2017 GPR impedance inversion for imaging and characterization of buried archaeological remains: a case study at Mudu city cite in Suzhou, China *J. Appl. Geophys.* **148** 226–233

[78] Halima A and Noor-E-A P 2018 The methods and recent invented tools and techniques used in archaeology for delicately preserving the past for the future *Archaeol. Discovery* **6** 338–54

[79] Lualdi M and Zanzi L 2004 2D and 3D experiments to explore the potential benefit of GPR investigations in planning the mining activity of a limestone quarry *Proc. 10th Int. Conf. on Ground Penetrating Radar GPR (June 21–24, Delft)* (Piscataway, NJ: IEEE) pp 613–6 https://ieeexplore.ieee.org/document/1343540

[80] Conyers L and Leckebusch J 2010 Geophysical archaeology research agendas for the future: some ground-penetrating radar examples *Archaeol. Prospect.* **17** 117–23

[81] Pérez-Gracia V, Caselles O, Clapés J, Osorio R, Canas J A and Pujades L G 2009 Radar exploration applied to historical buildings: a case study of the Marques de Lló palace, in Barcelona (Spain) *Eng. Fail. Anal.* **16** 1039–50

[82] Binda L, Cardani G and Zanzi L 2010 Nondestructive testing evaluation of drying process in fooded full-scale masonry walls *J. Perform. Constr. Fac.* **24** 473–83

[83] Cosentino P L, Capizzi P, Martorana R, Messina P and Schiavone S 2011 From geophysics to microgeophysics for engineering and cultural heritage *Int. J. Geophys.* **2011** 428412

[84] Lowe K M, Wallis L A, Pardoe C, Marwick B, Clarkson C J, Manne T, Smith M A and Fullagar R 2014 Ground-penetrating radar and burial practices in western Arnhem Land, Australia *Archaeol. Ocean.* **49** 148–57

[85] Urs B and Jens T 2010 Integrated Data Analysis At An Archaeological Site: A Case Study Using 3D GPR, magnetic, and high-resolution topographic data *Geophysics* **75** 169–76

[86] Giovanni L 2019 Principles of Mathematics Used in NDT Methods *Nondestructive Testing for Archaeology and Cultural Heritage: A Practical Guide and New Perspectives* (Cham: Springer Nature) p 1

[87] Valle S and Zanzi L 1998 Traveltime radar tomography for NDT on masonry and concrete structures *Eur. J. Environ. Eng. Geophys.* **2** 229–46

[88] Aleksandar R, Miro G, Lara P, Milan V and Željko B 2020 Using ground penetrating radar to reveal hidden archaeology: the case study of the Württemberg-Stambol Gate in Belgrade (Serbia) *Sensors* **20** 607

[89] Reynolds J M 1997 *An Introduction to Applied and Environmental Geophysics* (Chichester: Wiley)

Chapter 4

X-ray fluorescence spectroscopy in the investigation of works of art

Wael Badawy, Octavian G Duliu, Mina Magdy, Qunoot Mohammed Hadi and Sister Serafima

With the development of portable, handheld energy-dispersive x-ray fluorescence (EDXRF) spectrometers, this noninvasive, contactless analytical technique has demonstrated its remarkable advantages in the investigation of many types of artwork, from Paleolithic cave wall paintings to the most recent human artifacts. EDXRF is well suited to the simultaneous identification of more than 70 elements that are the main constituents of any material with a sensitivity reaching the order of 1–10 mg kg^{-1}. Together with Fourier transform IR (FTIR) and Fourier transform Raman (FTR) spectroscopy techniques, EDXRF can provide an almost complete characterization of the pigments, binders, and substrates of a great variety of works of art. This chapter presents the basic principle of x-ray fluorescence spectroscopy as well as representative examples in which EDXRF has proved its utility.

4.1 Introduction

Painted artworks are of great importance in art history and in archaeological contexts and therefore they are increasingly employed in scientific studies that involve investigations of materials [1, 2]. Colorants play a crucial role in answering technical and historical questions involving the manufacturing technique and provenance, indirect dating (*terminus ante quem*), the development of painting traditions, the trade in colorants, conservation strategy, and restoration intervention [3]. The colorants employed in paint manufacture are classified according to a variety of criteria, such as their composition (inorganic or organic), origin (natural or synthetic), or other characteristics. Pigments are inorganic fine solid powders that are insoluble in a binding medium (which can be beeswax, egg yolk, linseed oil, or acrylic resins) [4].

doi:10.1088/978-0-7503-4846-1ch4

Dyes are soluble organic substances that can be processed into insoluble pigments by precipitation or adsorption onto a metallic salt or an inorganic substrate; hence, they are called lake pigments [5]. The colorants can be derived from many sources: animal (indigo, madder, and charcoal black) or plant (cochineal and bone black) sources [6]. The colorants encompass a large variety of hues, and some of them can be obtained by mixing two or more colorants in suitable quantities [7].

Pigments have been used as coloring materials within artists' color palettes since prehistoric times [8, 9]. Pigment analysis can be complicated not only by the layered system of painting but also by the mixing of different pigments before painting (brushing) takes place or the degradation caused by prolonged exposure to environmental conditions such as light, humidity, and atmospheric gases [10]. Pigment identification is not based on color and hue criteria, so scientific examination is indispensable for their differentiation [11]. Pigments are either naturally occurring or artificially synthesized; for example, natural cinnabar and artificial vermillion are pigments that have the same chemical composition [7]. The pigments used in the practice of painting cover a broad range (table 4.1) [10, 12, 13]. It is worth noting that the colors of pigments are specific to the molecular arrangements of compounds rather than the constituent elements of the pigments. For instance, malachite, azurite, and verdigris pigments all contain copper in their chemical structures, but in different proportions and different crystalline structures and in combination with a variety of anions [14]. The choice of pigments is particularly crucial in order to achieve compatibility with the binder system, such as tempera (egg yolk), oil (linseed oil), tempera grassa (linseed oil and egg yolk), gouache (plant gum), encaustic (beeswax), and lime fresco (water) [4].

An archaeometric approach to pigment study is essential in order to establish any possible correlation between the chemical compositions of paintings and their historical context. This task can be accomplished by taking into account the history of each pigment, considering that from the second half of the 18th century, a significant number of inorganic pigments were synthesized and used for different periods of time (table 4.1).

For this reason, the correct identification of pigments can be very helpful in provenience studies. At the same time, given the uniqueness of artworks, the use of noninvasive techniques could be considered mandatory. As mentioned previously, in creating works of art, artists were never restricted to the use of a single type of pigment. The mixture of multiple pigments with binders created complex compositions, the thorough investigation of which requires a multitude of complementary analytical techniques. As the majority of pigments used since antiquity consist of various inorganic materials which are either natural or synthetic, the most appropriate noninvasive analytical technique appears to be x-ray fluorescence (XRF) [15]. This method, which is completely noninvasive, permits the correct identification of pigments, and, as the use of natural and man-made pigments has a long history which begins in the Upper Paleolithic (as demonstrated at Altamira, Spain [16], Lascaux, France [17], and Tassili n'Ajjer, Algeria [18]), it provides evidence of possible retouching, repairs, restorations, or forgeries.

Table 4.1. Categories of pigments used in the practice of painting according to their chemical composition.

Element	Pigment	Color	Chemical composition	When it was used
Arsenic (As) $Z = 33$	Orpiment	Bright lemon yellow to orange	As_4S_4	Since the 16th century
	Realgar	Bright orange	α-As_4S_4	From the 16th century until the middle of the 18th century
Antimony (Sb) $Z = 51$	Antimony black	Black	Sb_2S_3	Since 1500
	Antimony white (Timonox)	White	Sb_2O_3	Since 1920
	Naples yellow	Lemon to grayish yellow	$Pb_2Sb_2O_7$	Ancient Egypt to about 1850
	Titanium yellow	Yellow	$NiO{\cdot}Sb_2O_3{\cdot}20TiO_2$	Since 1954
Barium (Ba) $Z = 56$	Barite/blanc fixe	White	$BaSO_4$	Since 1820–1830
	Han blue	Blue	$BaCuSi_4O_{10}$	Ancient China, 1045 BC to 240 AD
	Han purple	Dark blue	$BaCuSi_2O_6$	
	Lemon yellow	Yellow	$BaCrO_4$	Early 9th century
	Lithopone	White	$BaSO_4{\cdot}ZnS$	Since 1870
Cadmium (Cd) $Z = 48$	Cadmium red	Orange to dark maroon	CdH_4SSe	Since 1907
	Cadmium yellow	Light yellow to orange	CdS	Since the mid-1840s
	Cadmium lithopone	Yellow	$BaCdO_4S_3Zn$	Since 1926
Calcium (Ca) $Z = 20$	Chalk	White	$CaCO_3$	Since the Neolithic period
	Gypsum	White	$CaSO_4{\cdot}2H_2O$	Since the 6th millennium BC
	Anhydrite	White	$CaSO_4$	Ancient Egypt
	Bone black/Ivory black	Black	C (10%) $CaCO_3$ and $Ca_3(PO_4)_2$	Since the 17th century
Chromium (Cr) $Z = 24$	Chrome yellow	Yellow	$PbCrO_4$	Since the 1820s
	Chrome orange	Pale yellow to deep orange	Pb_2CrO_5	Since 1809
	Chrome red	Intense red	$PbCrO_4{\cdot}Pb(OH)_2$	From 1809 until the beginning of the 20th century
	Strontium yellow	Pale yellow	$SrCrO_4$	Late 19th century
	Viridian	Blue–green	$Cr_2H_2O_4$ $Cr_2O_3{\cdot}2H_2O$	Since 1860

(*Continued*)

Table 4.1. (*Continued*)

Element	Pigment	Color	Chemical composition	When it was used
Cobalt (Co) $Z = 27$	Cerulean blue	Blue to dark blue	Co_2SnO_4	Since 1860
	Cobalt blue	Intense blue	Al_2CoO_4	Since 1806/1807
	Cobalt green	From intense green ($x \leqslant 0.3$) to pink ($x \geqslant 0.7$)	$Zn_{1-x}Co_xO$	Since 1835
	Cobalt violet	Intense violet	$Co_3(PO_4)_2$	Since 1859
	Cobalt yellow	Yellow	$K_3[Co(NO_2)_6]$	Since 1848
	Smalt	Blue	Potassium-rich cobalt glass	Since the second half of the 15th century
Copper (Cu) $Z = 29$	Atacamite	Soft muted green to emerald and blackish green	$CuCl_2{\cdot}3Cu(OH)_2$	Since antiquity
	Azurite	Deep to lighter blue	$2CuCO{\cdot}Cu(OH)_2$	Since antiquity
	Brochantite	Green	$Cu_4(OH)_6SO_4$	Since the 16th century
	Chrysocolla	Blue–green	$CuSiO_3{\cdot}nH_2O$	Since 3000 BC
	Conichalcite	Green	$CaCu(AsO_4)(OH)$	Since antiquity
	Egyptian blue, Pompeian blue, Cuprorivaite	Royal blue	$CaCuSi_4O_{10}$	Since 3000 BC
	Emerald green, Paris green	Pale blue to deeper green	$Cu(CH_3CO_2)_2{\cdot}3Cu(AsO_2)_2$	Since 1814
	Malachite verditer	Dark to lighter more transparent green	$Cu_2CO_3{\cdot}Cu(OH)_2$	Since the 15th century
	Posnjakite	Blue–green	$CuSO_4{\cdot}3Cu(OH)_2{\cdot}H_2O$	The late 18th century
	Scheele's green	Green	$AsCuHO_3$	Since 1775
	Verdigris/verdete	Dark bluish green	$Cu(CH_3CO_2)_2{\cdot}nCu(OH)_2$	Since antiquity
Iron (Fe) $Z = 26$	Iron earth pigments			
	Purple ochre	Intense red	Fe_2O_3 + clay (larger aggregates)	Since antiquity
	Red ochre	Reddish	Fe_2O_3 + clay (smaller aggregates)	Since the Paleolithic period
	Yellow ochre	Brownish yellow	Limonite + clay	Since the Paleolithic period
	Sienna	Yellowish brown	Limonite + MnO_2	Since the Renaissance

	Raw umber	Moderate yellowish brown	$Fe(OH)_3 + MnO_2$	Since the Neolithic period
	Burnt umber	Dark brown with a reddish tone	$Fe_2O_3 + MnO_2$	Since antiquity
	Limonite	Yellowish brown to light brown	A mixture of goethite, hematite, hisingerite, jarosite, maghemite, and lepidocrocite	Since the Paleolithic period
	Palygorskite/attapulgite	White, grayish, yellowish, gray–green	$(Mg,Al)_2Si_4O_{10}(OH)\cdot 4(H_2O)$	Pre-Columbian South America; mixed with indigo, it makes Maya blue
	Mars black	Warm brownish black	Fe_3O_4	Since the beginning of the 20th century
	Hercynite	Black	$FeAl_2O_4$	Since antiquity
	Prussian blue	Vivid blue	$(Fe_4[Fe(CN)_6]_3)\cdot xH_2O$	Since 1704
	Vivianite	Deep blue to deep bluish green	$Fe_3(PO_4)_2\cdot 8H_2O$	Since the 13th century
Lead (Pb) $Z = 82$	Anglesite	Lustrous white	$PbSO_4$	Since the Middle Ages
	Lead white	Soft white	$2PbCO_3\cdot Pb(OH)_2$	From antiquity until the middle of the 20th century
	Litharge	Yellowish orange	PbO (tetragonal)	Since antiquity
	Massicot	Red	PbO (orthorhombic)	Since antiquity
	Red lead/minium	Reddish orange	Pb_3O_4	Since antiquity
	Lead sulfide	Black	PbS	Since antiquity as a black glaze for pottery
	Lead–tin yellow (type I)	Canary yellow	Pb_2SnO_4	From the beginning of 14th century until the 17th century
	Lead–tin yellow (type I)	Saturated yellow	$Pb(Sn,Si)O_3$	From the 13th century until the 18th century
Manganese (Mn) $Z = 25$	Manganese blue	Clear azure blue to emerald green	$BaSO_4\cdot Ba_3(MnO_4)_2$	From 1869 until 1990
	Manganese violet	Deep rich violet	$NH_4MnP_2O_7$	Since 1868
	Pyrolusite	Brownish black	MnO_2	Since antiquity

Table 4.1. (*Continued*)

Element	Pigment	Color	Chemical composition	When it was used
Mercury (Hg) $Z = 80$	Cinnabar vermillion (synthetic cinnabar)	Brilliant red	HgS	Since antiquity
Nickel (Ni) $Z = 28$	Titanium yellow	Yellow	$NiO \cdot Sb_2O_3 \cdot 20TiO_2$	Since 1954
Silicon (Si) $Z = 14$	Glauconite/green earth	Green-grayish	$(K,Na)(Fe,Al,Mg)_2(Si,Al)_4O_{10}(OH)_2$	Since antiquity
	Phlogopite	Yellow to green	$KMg_3AlSi_3O_{10}(F,OH)_2$	Since the 19th century
	Lazurite (lapis lazuli)	Vivid blue	$Na_{8-10}Al_6Si_6O_{24}S_{2-4}$	Since the 13th century
	Ultramarine (Synthetic)			Since 1838
	Riebeckite	Black, dark blue	$Na_2(Fe^{2+}{}_3Fe^{3+}{}_2)Si_8O_{22}(OH)_2$	Since antiquity (Minoan frescoes)
Tin (Sn) $Z = 50$	Tin white	White	SnO_2	Since the 9th century, as a white glaze for pottery
Titanium (Ti) $Z = 22$	Titanium white	Intense white	TiO_2	Since 1918
Zinc (Zn) $Z = 30$	Zinc white	White	ZnO	Sporadically since the Middle Ages, intensively after 1834

More details of this technique and its application in investigations into many types of artwork will be further presented and discussed.

4.2 X-ray fluorescence

XRF is an x-ray emission spectroscopy technique that is widely used in the field of heritage science, since it provides information about the elemental composition of paintings, manuscripts, ceramics, bone, enamel, ivory, glass, jewels, coins, or any other objects of art [15, 19].

The use of XRF in elemental analysis is a consequence of Moseley's law, which states that the energy ϵ of an x-ray emitted by an atom, i.e. its characteristic radiation, is proportional to the atomic number, Z, squared, according to the equation:

$$\in = aR(Z^2 - b), \tag{4.1}$$

where a and b are two constants whose values depend on the element's period and R represents the Rydberg constant, which equals 13.606 eV.

Characteristic x-ray radiation appears when energy is transferred to an atom by irradiating it with low-energy photons (up to 50–80 keV) or electrons. As a result of this process, electrons are ejected from the lower shells, i.e. the K, L, M, or N shells, and their places are taken by electrons from superior shells. During this process, the electrons emit characteristic radiation whose energy is equal to the difference in energy between the levels involved (figures 4.1(a) and (b)). As the lower electronic shells are involved in this process, the energy of the characteristic radiation of elements whose atomic numbers are greater than six corresponds to the soft to moderate x-ray spectrum, i.e. between 2 and 85 keV.

Compared to optical or mass spectroscopy, XRF has three major advantages: (i) materials can be analyzed without any previous chemical treatment, such as acid digestion; (ii) the x-ray emission spectrum consists of a reduced number of lines, the so called x-ray characteristic radiation, whose energy monotonously increases with

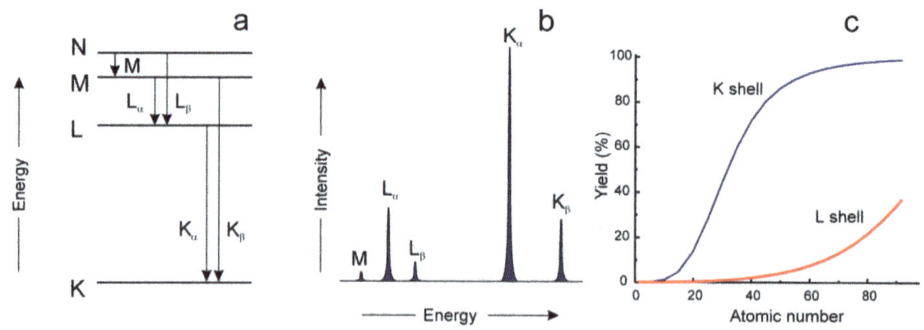

Figure 4.1. A schematic representation of the atomic shells involved in the emission of characteristic x-ray radiation: (a) the corresponding energy spectrum; (b) the atomic number dependence of the fluorescent yield. (c) The energies of the characteristic x-ray lines shown in (b) correspond to the distances between the electronic shells shown in (a).

the atomic number, as mentioned above; (iii) the energy of the characteristic radiation lies within the domain of soft to medium-energy x-rays, which is few orders of magnitude greater than the binding energy of atoms which forms molecules. This fact makes the characteristic radiation almost insensitive to the chemical combination, since the investigated materials manifest themselves like a mixture of isolated atoms regardless of their chemical combination.

As the energy spectrum of the characteristic radiation is specific to each element, this unique property permits its rapid identification. Moreover, the intensity of XRF characteristic radiation is, to a first approximation, proportional to the mass fraction of the corresponding element, which therefore allows a semi-quantitative to fully quantitative elemental analysis to be obtained.

The fluorescent yield η represents a parameter which quantifies the emission of characteristic radiation. The fluorescent yield is defined as the ratio between the number of vacancies which are filled by upper-shell electrons (accompanied by the emission of characteristic radiation) and the total number of vacancies. It is better described by the following equation [20]:

$$\eta = \frac{Z^4}{A + Z^4}, \tag{4.2}$$

where A is a semiempirical parameter equal to about 10^6 for the K shell and 10^8 for the L shell (figure 4.2(c)).

For this reason, the intensities of the L lines are always smaller than those of the corresponding K lines (figure 4.1(b)).

In this regard, figure 4.2 illustrates typical XRF spectra of some dyes and primers used currently. All the paints were prepared in tempera, while the primer was prepared from chalk and bone glue and partially covered with varnish consisting of

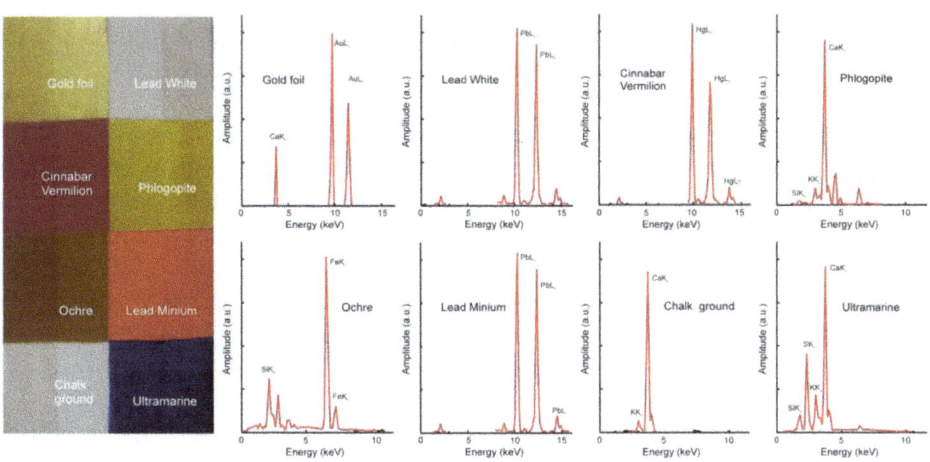

Figure 4.2. Typical XRF spectra of some pigments and binders currently utilized in easel and wood painting. Reproduced from [21] Copyright © 2019 Romanian Academy Publishing House. All rights reserved.

a mixture of damar and turpentine. It is worth noting the absence of Sr lines in the XRF primer spectrum, as Sr usually accompanies calcium in gypsum.

Although XRF is a well-known analytical method whose principle is apparently simple, routine measurements should take into account and correct for the influence of neighboring atoms. Depending on the atomic number and mass fraction, these atoms may either partially absorb the characteristic radiation emitted by a given element, or, if their characteristic radiation has a higher energy, supplementarily stimulate the emission of the characteristic radiation of the atoms of interest.

These effects, known as matrix effects, can significantly alter the proportionality between the mass fraction of the elements of interest and the intensity of their characteristic x-ray radiation. The absorption effect can lead to apparently smaller values of the mass fractions of the elements of interest, while the enhancement effect can result in apparently higher values of the mass fractions.

Depending on the complexity of the analyzed material and the number of different elements it is composed of, both effects can manifest themselves simultaneously, ultimately affecting the precision and accuracy of measurements. At the same time, in the case of quantitative XRF, the polychromatic energy spectrum of the actual x-ray generators constitutes a supplementary source of errors.

Given these peculiarities, the intensity I_i of the characteristic line of element i in a sample containing N different elements can be correlated with its mass fraction c_i by the following quasi-empirical equation [20]:

$$I_i = c_i \frac{1 + \sum_{j=1}^{N} e_{ij} c_j}{1 + \sum_{j=1}^{N} a_{ij} c_j}, \tag{4.3}$$

where e_{ij} and a_{ij} represent the enhancement and absorption contributions, respectively, of each element contained within the investigated object and whose values are obtained from a set of appropriate certified reference materials (CRMs).

This complex system of equations, as a rule, can solved by dedicated software, majority of them implemented into actual XRF spectrometers

Another peculiarity of the XRF analytical method is related to the relatively shallow layer of analyzed material. Indeed, the analyzed volume should be thin enough to permit the characteristic radiation to reach the detector (figure 4.2) in the case of total reflection x-ray fluorescence (TRXRF). In this setup, the angles of incidence and reflection angle are approximately 45°, which enhances the incident area of the x-ray beam, resulting in lower penetration depth, reduced background noise, and the simultaneous diminution of the matrix effects—ultimately leading to higher sensitivity [22].

In the case of heritage materials, another drawback of XRF investigation is related to the stratification, heterogeneity, contamination, or degradation of object surfaces, which, as in the case of the analysis of complex samples, recommends the use of portable XRF spectrometers [23].

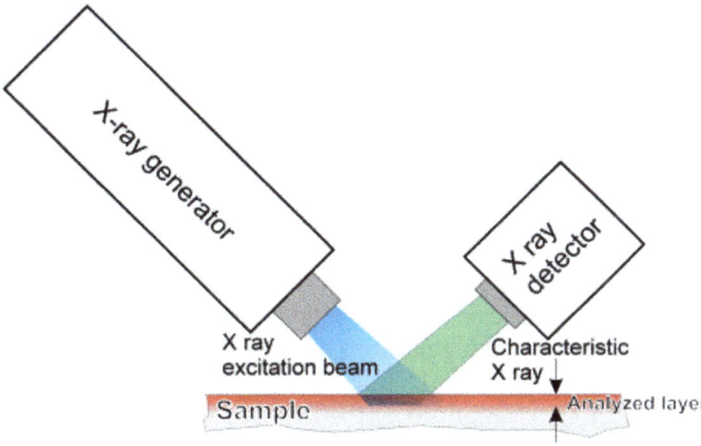

Figure 4.3. Diagram of the arrangement of an x-ray generator, the sample, and an x-ray detector in a classical TRXRF setup. Due to the attenuation of both the incident and characteristic radiation, the analyzed volume is reduced to a shallow layer under the object's surface. In the case of macro x-ray fluorescence (MA-XRF) mapping, one or both x-ray beams can be localized to the object's surface using poly-capillary collimators [24].

The main information related to the elemental composition of a work-of-art material or object, regardless of its nature, e.g. paint layers, metallic items, ceramics and ceramic paints, paper, parchment, etc. is furnished by its XRF spectrum, which, as a rule, consists of a superposition of several characteristic radiation lines (figure 4.3). Under these circumstances, the higher the energy resolution of the x-ray detector, the better the individual lines of the characteristic radiation are evidenced. At present, two main types of detector are used to record and analyze the energy spectrums of characteristic radiation: energy-dispersive XRF (EDXRF) and wavelength-dispersive XRF (WDXRF) [20, 23].

The WDXRF method has better spectral resolution, but needs complex x-ray optics, and, as the spectrum is recorded sequentially, a longer acquisition time. For this reason, WDXRF cannot be completely noninvasive, as small pieces of investigated material must be finely ground, mixed with an inert material such as lithium tetraborate or another low-atomic-number binder, pressed into small pellets and exposed to x-rays. To reduce the air attenuation of the characteristic radiation, all measurements should be performed in a vacuum chamber.

In contrast to WDXRF, EDXRF [25] has better efficiency due to the use of semiconductor detectors that can quasi-simultaneously record x-ray photons with different energies [20], but at a slightly lower energy resolution. This relative smaller disadvantage is compensated by the compactness of such devices which, as handheld instruments, have been shown to be extremely useful in field investigations.

As mentioned above, the energy of characteristic x-rays is independent of the chemistry of the material, so the XRF method cannot be used to distinguish between different salts of the same element. Another limitation of this technique is related to interference between the K and L lines of different elements, as in the case of barium ($L_\alpha = 4.46$ keV) and Ti ($K_\alpha = 4.50$ keV), whose energy difference [26] is smaller than the energy resolution of the actual silicon P intrinsic N (PIN) diode or SDD (silicon drift detector) [27].

With the exception of coins and jewelry, most works of art have sizes from centimeters to meters. As they are displayed in art galleries, museums, churches, temples, or private collections, they are difficult or almost impossible to transport to a laboratory for examination and investigation. Under these circumstances, the use of mobile XRF spectrometers seems the best solution, especially as these are portable instruments equipped with miniature x-ray generators and PIN or SSD detectors cooled by the Peltier effect (the best resolution being attained by the SDD-based instruments). As in the case of other solid-state x-ray detectors, SSDs measure the amount of energy the incoming photons transfer to the detector material through ionization, because the electrical charge thus generated is proportional to the transferred energy. In the case of SSDs, the detector itself consists of a high-purity silicon disc, about 0.5 mm thick, with a very low leakage current due to supplementary cooling by the Peltier effect. The electric charge produced by incident x-rays drifts under the action of a radial electric field generated by a series of coaxial ring electrodes (placed on the opposite face to the incident radiation detector face) until is collected by a small anode placed at its center. From here, the collected charge is amplified by a field-effect transistor (FET), which generates an electric impulse whose amplitude is proportional to the released energy. Due to their planar configuration, SSDs have a very low capacitance, allowing a high count rate of up to 10^6 cps and an energy resolution of about 125–128 eV for the Mn K_α line [27].

On the other hand, the use of silicon, a low-atomic-number material (Z = 14) restrains the range of x-ray energies to 15–20 keV, which corresponds to the Lγ line of lead (E = 14.76 keV) [25]. Therefore, the presence of heavy elements such as Hg, Au, or Pb is measured using the L lines of their XRF spectrums. By contrast, due to the low yield of elements whose atomic numbers are lower than 11 (for example, the $K_{\beta 1}$ line of sodium has an energy of 1.07 keV), the use of portable spectrometers in such cases is very limited. This difficulty is exacerbated by the attenuation of the ultra-low-energy characteristic radiation by the air and detector windows. To overcome this inconvenience, such measurements should be made either in vacuum or by purging the space between the object and the detector with helium.

For this reason, although XRF is very useful in identifying inorganic metal-based substrates or mordants, it cannot be used to identify organic dyes or lakes or any other carbon-based materials. When used to investigate paintings, EDXRF is not always able to discriminate between information from different paint layers, although if a layer containing light elements covers a layer consisting of heavy elements such as Hg and Pb, their presence can be still detected.

The small size of SSD detectors, together with the miniaturization of x-ray generators, has allowed the creation of miniature, handheld XRF spectrometers. These instruments, due to their performance and simplicity of use, are currently used *in situ* to investigate a great diversity of works of art and archaeological artifacts. Unfortunately, such instruments do not offer a vacuum path between the sample and the analyzer, limiting their use to materials that have atomic numbers greater than 11 (i.e. that of sodium), although, as mentioned above, the use of a helium purge can lower this limit to that of fluorine.

To increase the performances of XRF applications in the study of works of art, in addition to more mobile handheld XRF spectrometers, the development of x-ray optics such as poly-capillary collimators has permitted the scanning of the entire surfaces of paintings or other objects. Known as macro XRF (MA-XRF) mapping [24], this technique generates maps that portray the distributions of different elements present in a pigment. By comparing the resulting XRF maps with optical images, it is possible to evidence the pigments or pigment combinations used with more accuracy, permitting a better understanding of the painter's technique.

It is worth mentioning that XRF plays a major role in our expertise in works of art, due to its totally noninvasive character. At the same time, its inability to be used in investigation of organic matter limits its utilization to some extent; thus, for more detailed characterization, a complementary spectroscopic method such as FTIR or FTR spectroscopy is used. As a result of instrument miniaturization, both techniques can be used in their noninvasive, contactless variants such as external reflection FTIR (ER-FTIR) [28] and FTR spectroscopy [29–31]. To these could be added micro-invasive methods such as organic mass spectrometry [32], which combined with ER-FTIR or FTR permits the identification of various organic moieties as well as inorganic pigments. The disadvantage of XRF in terms of its difficulty in evidencing low-atomic-number elements can be overcome by using accelerated protons to generate the emission of characteristic x-ray radiation, a method known as particle-induced X-ray emission (PIXE) [33]. Although complementary to XRF, PIXE is a microinvasive technique, which limits its applicability. In a few cases, XRF measurements have been accompanied by x-ray diffraction (XRD) measurements, which permitted the unambiguous identification of the minerals that were included in a pigment's composition [34]. In addition to these, classical or digital x-ray radiography has proved to be an excellent auxiliary technique and has been intensively used to reveal hidden details such as overpainting [35], restorations [36], and other defects [37].

As mentioned above, EDXRF has shown great potential for the investigation of heritage objects. EDXRF spectra give immediate information about the natures of almost all materials, provided that their atomic numbers are, as a rule, greater than 11. A quick glance at the chemical compositions of the most utilized pigments (table 4.1) demonstrates that all of them can be identified by XRF, which is reflected by the impressive volume of publications dealing with paintings from Fayum portraits [38] to 20th century Russian avant-garde works and even more recent work [39, 40].

Depending on the interest or on the available instruments, EDXRF investigations can be done locally by choosing the most representative details, which is the case in the majority of studies (figure 4.4); alternatively, or the MA-XRF [24] technique can be used to scan the painting's surface, which results in full maps that portray the distribution of pigments over the entire painted surface [41, 42] (figure 4.5).

More detailed information about the paint layer can be obtained in a micro-invasive mode by means of scanning electron microscopy energy-dispersive XRF (SEM-EDX). In this case, a small fragment of the paint layer, which can include the

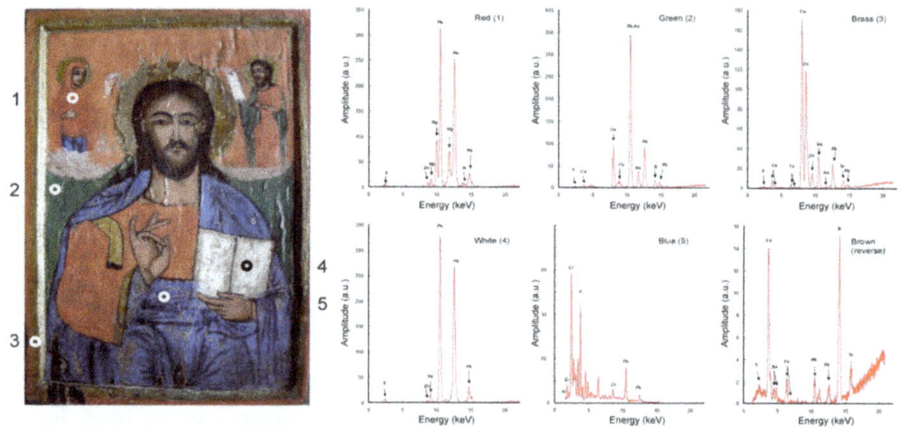

Figure 4.4. The icon of Jesus Christ as the Righteous Judge (end of the 18th century) and some of the most representative XRF spectra of red (lead minium and vermillion), green (probably verdigris and orpiment), brass powder (a replacement for gold), white (lead white), blue (most probably ultramarine), as well as the brown on the reverse dominated by gypsum, ochre, and lead white, presumably pigments. The intense K_α line of lead appears almost everywhere, suggesting the use of a lead-rich white primer. Reproduced from [43] Copyright © 2010 Romanian Academy Publishing House. All rights reserved.

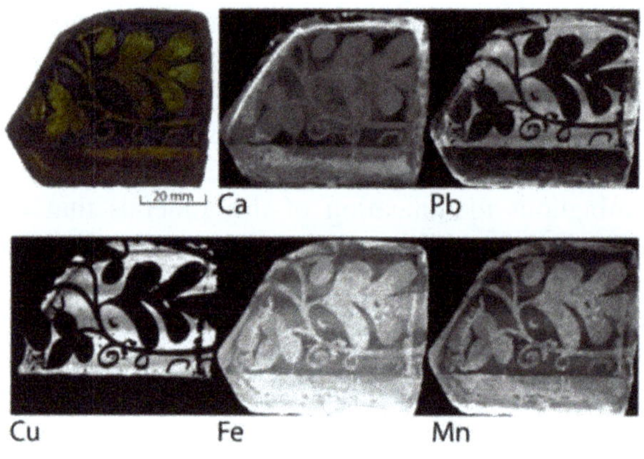

Figure 4.5. MA-XRF image of a glass pane belonging to a stained glass window and exposed at the Diocesan Museum in Kielce, illustrating the distributions of calcium, lead, copper, iron, and manganese of which the pigment is composed. Reproduced from [41] CC BY 4.0 © 2021 by the authors. Licensee MDPI, Basel, Switzerland.

primer and fragments of the substrate is embedded in resin, cross-sectioned, polished, and then examined, which allows the elemental compositions of the individual layers to be separately identified by EDXRF. Moreover, as in the case of MA-XRF, SEM-EDX furnishes a map depicting the 2D distribution of the elements used in the primer and pigments (figure 4.6) [44]. Such maps yield valuable information about the techniques used by each painter or painting school, while at the same time providing a valuable resource that can be used in planning for any future restoration work.

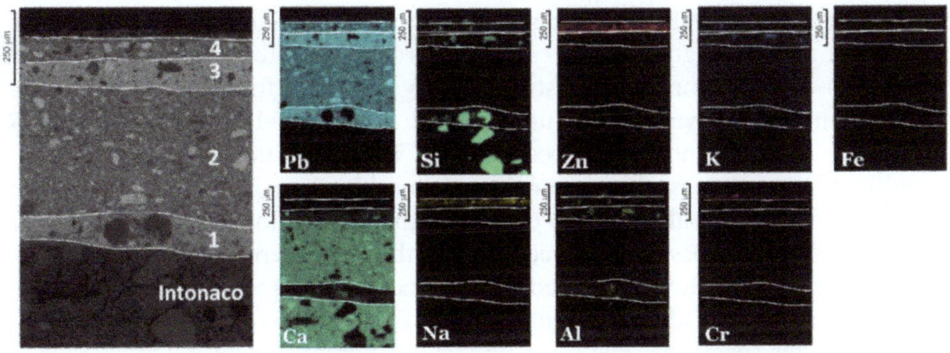

Figure 4.6. A backscattered electron SEM image of a cross section (left image) illustrating the priming coat applied to the 'intonacio' layer. This image was necessary for the restoration of a 19th century Belgian wall painting; it confirms the presence of four priming layers, all of them rich in lead, which suggests the use of a lead-rich white primer. The Si and Ca are consistent with the use of quartz and calcium carbonate, while Mg, Al, K, and Fe are consistent with the presence of green earth observed under optical microscopy. Reproduced from [42] CC BY 4.0 © 2021 by the authors. Licensee MDPI, Basel, Switzerland.

4.3 Applications

4.3.1 Rock and wall paintings

Rock paintings are among the oldest human artifacts which have been studied by XRF combined, in some cases, with other complementary methods, such as FTIR and FTR spectroscopy.

In this regard, EDXRD data obtained from 56 sampling points at La Peña Cave in San Roman de Candamo (Asturias, Spain) showed that Si correlated at a Pearson's r coefficient greater than 0.8 with Al and Fe. This finding was explained by FTR observations suggesting the use of natural iron oxide aluminosilicates. At the same time, in spite of the presence of a natural carbonate substrate, Ca showed a negative correlation with Fe at an r value of about −0.90, a fact explained by the thicker pictorial layer which was rich in Fe—a strong absorber of x-rays—which significantly attenuated the Ca signal of the substrate [44].

In a similar study performed on an ancient rock painting in Erongo, Namibia using a portable XRF spectrometer, it was shown that iron oxide, carbonaceous black, apatite, and calcium sulfate were used by the ancient artist. Although significant portions of the painted surfaces were altered, this kind of study has been extended to paintings other than the already-known paleolithic wall paintings [45].

The aboriginal wall paintings found in various regions of Australia have also been investigated in detail using handheld XRF instruments. This showed that in the majority of cases, the pigments used consisted not only of iron oxides and hydroxides but also of manganese oxide and charcoal black [46].

Fresco is a technique of mural painting employed since antiquity, but it is mainly associated with the Italian Renaissance. Regardless of the variant used, i.e. fresco or fresco secco, this technique consists of applying pigments to a thin layer of lime plaster, so that the pigments merge with the plaster and in this way become an integral part of the wall. Since, until the 20th century, fresco paintings mainly consisted of mineral pigments, XRF is an appropriate method for the study of the pigments as well as the plaster substrate. A few examples follow.

Pre-Columbian Maya wall paintings were investigated in detail using XRF combined with FTR and FTIR spectroscopy. This permitted the identification of a wealth of pigments, such as cinnabar, hematite, ilmenite, Maya blue, ochre, limonite, and goethite, all of which explained the vivid color of the paintings [47].

The Knossos frescoes, considered among the most representative of Minoan culture, were thoroughly investigated by analytical techniques, among which, XRF was helpful in pigment identification. This was done *in situ* following subtraction of the XRF plaster spectrum. The most important pigments were Egyptian blue and riebeckite (table 4.1), which were used as blue colorants, red and yellow ochre, and lime, which was used as a white colorant. A complementary XRD analysis evidenced the presence of hematite, illite, glauconite, as well as goethite, i.e. minerals of local origin. A black pigment was identified as charcoal [48].

With the advent of the Roman Empire, the use of painted plaster to adorn public edifices such as baths was extended to the entire Imperial territory. This was the case for the Roman province of Dacia, where fragments of mural painting were discovered at the Alburnus Maior, the actual Rosia Montana archaeological site located in an older Roman mining center for gold extraction. Here, wall painting fragments were investigated using multiple noninvasive techniques including EDXRF [49]. This permitted the identification of Ca as a major element together with traces of Sr as well as Ti, Mn, Fe, Cu, Zn, As, and Hg, all of which suggest the use of red, yellow, and manganese-rich ochre, i.e. mainly *austeri* pigments (common earth pigments) [50].

The same technique was utilized to investigate the terracotta statue of *Zeus Enthroned* from the ancient city of *Poseidonia*, known also as *Prestum*, located in the plain of the Sele river near Salerno, Italy. In addition to XRF, SEM-EDX, XRD, and FTR spectroscopy were used for a better characterization of the terracotta material as well as the paint layer. According to the XDR results, a calcareous clay rich in chlorite was used for the statue's body; this consisted of two clay layers fired under oxidizing conditions but at different temperatures. At the same time, the results confirmed the presence of two decorative layers: a brown one created by mixing iron oxides which, by fairing, transformed into hematite and jacobsite, and a red one, consisting of pure hematite. Hence, the pigments employed in the decoration technique suggested that the production technology used was outside the ancient Greek domain, and, at the same time, proved the existence of cultural and technological transfers between Greek colonies and Italic pre-Roman societies [51].

The restoration work at the Roman Theater Nuevo in Medellín, Spain, which transformed an ancient Roman site into a renowned cultural heritage building, needed a detailed investigation of some of the well-preserved parts of the ancient theater, such as the *proedria*, *orchestra*, *frons pulpiti*, *scaenae frons*, etc. The unearthed stucco columns and remains of original wall paintings were mainly analyzed *in situ* using EDXRF, and only small fragments of painting and stucco were analyzed in the laboratory by XRD, FTIR, and SEM-EDX spectroscopy. The main constituents of the studied objects were Ca and Fe with traces of Mn, suggesting the use of different iron oxides for red and ochre colors in addition to the use of calcium carbonate. For some pieces, the Mn was absent, meaning that umber pigment was not always used. Also remarkable was the total absence of Hg, the main constituent of cinnabar, in spite of the proximity of the Almadén cinnabar mine [52].

XRF was used to analyze the pigments used in the polychromatic painting of busts and coats of arms in the *triforium* of St. Vitus Cathedral, Prague. In this way, it was possible not only to identify the pigments, but also, given the periods when some of them were used, to evidence repainted sections. Here, as a general characteristic of the Medieval period, the pigments used were of a reduced diversity, mainly red lead, cinnabar, red ochre, and azurite as well as smalt, the last of which attests a 19th century repair, as smalt was used in Europe beginning in the second half of the 15th century (table 4.1) [53].

The same combination of XRF and FT Raman was helpful for an *in situ* investigation followed by a microinvasive laboratory study of the mural paintings of the 15th century Romanic church of Saint Ementerio and Saint Celedonio of Goikolexea (Basque Country, Spain), which were partially restored in the 17th and 20th centuries. In addition to the presence of gypsum, thenardite ($Na_2SO_4{\cdot}10H_2O$), and some nitrate compounds, e.g. nitratine ($NaNO_3$) and nitromagnesite ($Mg(NO_3)_2{\cdot}6H_2O$), the study found evidence for the presence of connellite, ($Cu_{19}(OH)_{32}(SO_4)Cl_4{\cdot}3H_2O$), a hydrous copper chloro-sulfate, all of which testify to a significant degradation of the original fresco due to the poor conservation conditions and the prolonged action of natural humidity. Concerning the rare mineral connellite, it is worth mentioning that its presence is due to the degradation of malachite, the only Cu-containing pigment utilized in the original painting [54].

To monitor restoration works, Giotto's frescoes in the Padua Chapel of the Scrovegni were systematically analyzed *in situ* using a portable XRF spectrometer provided with a Ca-anode x-ray generator working at 4.5 keV. This setup permitted the detection of the low-energy (up to 4 keV) characteristic x-ray radiation of S, while to detect other elements, the x-ray generator used a W anode at an anodic potential of 30–35 keV. In both cases, a Si PIN diode was used as the detector. Sulfur monitoring was necessary as the cleaning process was carried out using anionic ion-exchange resins. The same portable spectrometers were useful in an investigation of the golden and black auras, the former of which were made of pure gold, most probably obtained by hammering contemporary gold coins from Venice or Florence according to [55]. In the case of the black auras, XRF analysis indicated that Pb was the main component with small impurities of S, Ca, Fe, and Cu [56].

A portable XRF spectrometer was also successfully used for *in situ* analysis of pigments in twenty frescoes of the 16th century Assumption Cathedral of Sviyazhsk, Tatarstan Republic, Russian Federation. The results obtained from 193 noninvasive samplings finally indicated different recipes, the main components of which were MgO, CaO, Al_2O_3, SiO_2, P_2O_5, SO_3, Fe_2O_3, Sb_2O_3, and PbO, which permitted the inference of the most probable pigment recipes used by the Russian artisans.

The white pigment consisted of slaked as well as dolomitic lime mixed in various proportions. Antimony and bone white were used for the basic white batch, while red and occasionally blue ochre, green earth, realgar, orpiment, galena, antimony black, and magnetite-based pigments were also used in the paints [57].

To unequivocally determine the nature of the blue pigment which adorns the external mural paintings of a medieval church in Moldova, an *in situ* portable XRF spectrometry was combined with microinvasive FTIR spectroscopy; the results showed the presence of azurite and smalt, both pigments used in medieval wall painting in Romanian provinces. It is worth mentioning that the best results are obtained by combining EDXRF with FTIR or FTR spectroscopy, as the last two methods permit the identification of specific vibration modes of different moieties [58].

Portable, handheld XRF spectrometers were used to determine the nature of pigments used in wall paintings of the Chapel of Agios Georgios, Haidari, Athens, dated to the second half of the 19th century. Although this is an almost contemporary painting, XRF data were useful in evidencing the presence of various mineral pigments, e.g. malachite, cobalt blue, ochre, burnt Sienna, goethite, red and white lead, realgar, and bone black, the last of which was evidenced by the presence of Ca and traces of P. However, as in many cases, portable FTR spectroscopy finally permitted the pigments to be identified [59].

4.3.2 Ceramics

Pottery could be considered to be one of the oldest human creations. The oldest artifact, the Venus of Dolní Věstonice, Czech Republic, which most probably originated at the end of the Paleolithic period, dates back to 29 000–25 000 BC [59]. Initially, ceramic ware was unpainted, but by the end of the Neolithic period, i.e. about 5500 BC, the first fragments of painted pottery were being created (as evidenced by finds in the Mediterranean basin) [60].

Trichrome painted wares spread along the eastern side of Italian peninsula during the Middle Neolithic (between 5000 and 4500 BC). These artifacts consisted of a very fine granulated paste of light red-brownish color painted with red and black pigments. To establish the pigments' nature, further fragments of this type of ceramic, collected in Abruzzo, Central Italy and Basillicata, Southern Italy, were investigated using EDXRF and micro FTR spectroscopy. Both methods proved the presence of a black pigment based on manganese–iron oxide, while the red pigment mainly consisted of iron oxides [61].

Minoan pottery was also investigated using XRF in combination with XRD. In spite of the use of multiple sampling points, the analysis only documented the presence of iron pigments, e.g. hematite for red pigment, while a black color was obtained using a mixture of magnetite and hercynite. Brown and purple pigments were probably prepared by the mixture of these pigments with a clay rich in calcite; all of these materials, as in the case of Minoan frescoes, were of local provenience [62].

EDXRF was used to analyze the bichromatic decoration of two different types of ancient Cypriot pottery, i.e. Cypro-Geometric and Cypro-Archaic bichrome ceramics found in Tel Dor, Israel and dated to between 1200 and 500 BC. In this way, it was shown that the black paints consisted of manganese-based pigments comprising manganese and iron, while the red paints consisted of iron-based pigments rich in iron and poor in manganese. These results proved that the local Cypriot black umber and red ochre were for this kind of ceramic [61], which is similar to the findings for Minoan pottery [63].

Handheld EDXRF was used to evidence the elemental fingerprints of 26 fragments of painted ceramic ware dating from the Neolithic to the late Bronze Age and collected from the Kur river basin in the southwestern Zagros mountains of the Iranian Fars province. Principal component analysis (PCA) allowed the ceramics to be classified into two clusters, one of which was a group of five sites, while the other one was restricted to a single site, suggesting the existence of only two major sources of clay. In addition, EDXRF revealed the presence of an intense Mn K_α line, suggesting the use of Mn oxide in the preparation of a black pigment, a characteristic found in most ancient ceramics disseminated along the Mediterranean Sea basin [64].

Returning to the pre-Hispanic Mesoamerican Maya culture, a series of investigations into ceremonial funerary artifacts decorated with blue, red, and white pigments and belonging to the West Mexican Chupicuaro culture—600 to 100 BC was performed using EDXRF, SEM-EDX, FTR spectroscopy, and liquid chromatography quadrupole time of flight mass spectrometry (LC-MS/MS-TOF/MS). The results of these investigations confirmed the presence of hematite, cinnabar, and calcium carbonates as well as different species of indigo, e.g. *Indigofera suffruticosa* L., *I. mucrolata*, and *I. jamaicensis*, the main constituent of the organic–inorganic Maya blue pigment outside the traditional Maya region. This finding reconfirmed the existence of close contacts between different pre-Columbian American civilizations [65].

The simultaneous use of noninvasive as well as microinvasive XRF together with FTR spectroscopy was successfully used to determine the composition of a polychromatic glaze used by the 16th century renowned tile factory of Niculoso Francisco, nicknamed ‘Pisano.’ The results for two altarpieces of the chapel of Queen Isabella the Catholic (Royal Alcázar of Seville) and Our Lady of Tentudía in the homonymous monastery in Segura de León (Badajoz) documented the presence of lead antimonate (Naples yellow) in yellow and beige glazes, various proportions of Mn and Fe in brown sectors, Cu, Co, and Zn in the green and blue parts, as well as the presence of Sn oxide in the case of a white substrate. All these elements appeared in different proportions, testifying to the artist’s excellent craftsmanship in combining different pigments to achieve a wide range of colors and shades [66].

Microinvasive WDXRF and portable EDXRF were used to determine the elemental composition of 12 fragments of blue-and-white porcelains of Chinese Yuan and Ming Dynasties (1271–1644 AD). For a complete study, the body, glaze, and blue decor were investigated. The body’s elemental composition provided evidence that all samples belonged to a type of porcelain produced in southern China. The same finding was supported by the elemental composition of glaze, while the blue decor showed a variable Co content, the Mn/Co and Fe/Co ratios being good indexes for discrimination purposes [67].

EDXRF was used to evidence the influence of the Second World War (WWII) on the Mn/Co ratio of the Chinese porcelain blue glaze. Indeed, the blue glazes of Chinese porcelain objects manufactured before WWII had Mn/Co ratios that varied between 2.59 ± 0.20 and 4.93 ± 0.64, which suddenly decreased to a minimum of zero and a maximum of 0.41 ± 0.05, respectively, most probably due to a significant change in the supply chains. This makes the Mn/Co ratio an appropriate criterion for use in provenance studies of Chinese blue glazed porcelain made in the WWII period [68].

4.3.3 Sculptures

EDXRF, as an essentially noninvasive technique, was involved in identifying the pigments used to paint temples and statues in Classical Antiquity, as in the case of Athenian Acropolis. Although almost the entire paint layer had been washed out over time, it was still possible to evidence the use of azurite for light green–blue, Egyptian blue for vivid light blue, malachite and conichalcite for bright green, and hematite, red ochre, and cinnabar for red. The apparently reduced number of pigments is mainly due to the action of external factors such precipitation, air pollution, and wind, which, as mentioned above, had almost washed out the entire paint layer [69].

The use of synchrotron radiation, due to its high output and quasi-monochromatic character, coupled with an innovative color x-ray camera [70], permitted the reconstruction of the distribution of P, Ca, and Sr in a carved Phoenician ivory object dated back to the 8th century BC together with remnants of Cu, Mn, Fe, Ti, and Au. The Cu was considered as having originated from the former paint layer, Fe and Pb were most probably the debris of sediments or pigments, Mn came from microscopic fragments of soil minerals, Ti most probably originated from traces of previous restorations or soil debris, and Mn, Fe, and Au all attest to a former gilding [71].

Two wooden sculptures dated to the 16th–18th centuries were thoroughly investigated by combining EDXRF with digital radiography to evidence different restorations performed over time, including recent (20th century) ones. Carved in Portugal and transported to Brazil, both pieces are examples of sacred art and represent Saint Sebastian and Our Lady of Conception. The pigments evidenced in the paint as well as in the underpainting layers were specific to the 16th century with the exception of Prussian blue, which was utilized for the first time at the beginning of the 18th century. This finding suggests either that a restoration was performed after Prussian blue had started to be used in European painting or that the sculptures were indeed carved in the 18th century [72].

Glazed and unglazed terracotta sculptures have been known since antiquity, but from the time of the Italian Renaissance, these techniques gradually spread over the entirety of Western and Central Europe. To check the actual status of two Pietro Torrigiano sculptures from the collection of the Fine Arts Museum of Seville, noninvasive XRF was shown to be one of the most appropriate techniques for *in situ* investigation. This permitted the presence of major and some trace elements to be detected both in unpainted substrates and paint layers. The results evidenced the presence of Al, Si, K, K, Ca, Mn, and Fe as major elements, thus confirming that terracotta was the raw material, while traces of Ti and Zn revealed the presence of some modern retouching. Given the sculptures' age, XRF evidenced that umber, red ochre, cinnabar, azurite, and some copper-based green pigments as well as an organic black pigment, most probably carbon black, had been applied to the terracotta substrate [73].

Not only stone, marble, lime, wood, and terracotta have been investigated by XRF. Bronze, in its turn, has been intensively used since antiquity to create remarkable works of art. Compared to the other materials used in sculpture, bronze has a relatively stable composition that is mainly restricted to Cu and Sn with small impurities of other metals such as Al, Mn, Ni, and Zn as well as P, Si, or As. Bronze was a product of the homonymous era which started in western Eurasia and India in the mid-4th millennium BC. Since that time, an impressive number of bronze works of art were created, many of them surviving to the present day. The presence of Cu, Zn, and Sn and that of other heavy metals make XRF an appropriate technique for analysis. Therefore, noninvasive XRF was used together with XRD, scanning electron microscopy (SEM), and optical microscopy to obtain actual data for the elemental composition as well as the corrosion products of a late 19th century outdoor statue of the Portuguese king, Dom Afonso Henriques, the first king of Portugal. The results of this investigation evidenced a ternary alloy of Cu, Sn, and Zn with traces of Mn, Fe, As, Bi, and Pb, which was close to the bronze composition utilized by the same 19th century foundry [74].

Although XRF cannot be classified as a chronological method, it was used to determine the elemental structure of a Parisian cast gilt bronze with sufficient confidence; thus, by comparison with the continuous evolution of the French metallurgical craft, it permitted an evaluation of the real ages of bronze artifacts. This was possible due to the existence of a rich database of the bronze compositions from the 17th century to the early 20th century. The compositional data for 466 bronze objects of known age was aggregated, recalibrated, and used with support vector and random forest regression machine learning programs to estimate the ages of more French gilt bronzes with an accuracy and precision of 90%. These achievements proved once more the utility of XRF not only in determining the real ages of objects but also in evidencing forgeries [75].

4.3.4 Artifacts, cartonnage, parchments, manuscripts, and prints

Ancient art artifacts, regardless of their nature and age, are among the objects most intensively investigated using XRF, as in the case of an engraved dyed ivory artifact, a product of 8th century ancient traditional Japanese *bachiru* craftsmanship. In this technique, the ivory is stained by a repeated process of boiling and soaking until it reaches the desired hue; it is then repainted with different motifs. This technique was used to create the Shosoin treasures held at the Tōdai-ji Temple in Nara, Japan.

In this regard, EDXRF was used to elucidate the colorants present in these paintings. The corresponding XRF spectra evidenced the characteristic peaks of Cu, which, in connection with the XRD investigation of the same paint layer, allowed atacamite to be identified as the main green pigment. In addition to XRF, reflectance spectrometry was used to evidence sappanwood red dye, lac, madder, gromwell, and indigo as other organic components of the Shosoin treasures [76].

Egyptian cartonnage, i.e. the material used in ancient Egypt from about 2150 BC to the Roman era, is older than *bachiru* objects. Cartonnage was investigated by analytical techniques to evidence the nature of the pigments, binder, and supporting material used. The cartonnage consisted of one or more layers of papyrus or linen plaster painted in polychromy. This technique was used to realize funeral masks and even Fayum mummy portraits. In this regard, EDXRF, μFTIR, μRaman, SEM-EDS, and pyrolysis–gas chromatography–mass spectrometry (Py-GC/MS) (Py-GC/MS) were used for a comparative investigation of cartonnage artifacts stored in the Archaeological National Museum, Cairo, the Archaeological Museum of Lisbon, and the Natural History and Science Museum of the University of Porto, Portugal. In this way, it was found that in addition to the yellow and red ochre, realgar, cinnabar, and Egyptian blue and green used as pigments, the paint layer contained the unexpected presence of Sn, most probably associated with recycled bronze scraps used during pigment preparation [77].

A quite similar investigation examined some cartonnage fragments found in the collections of the Egyptian Museum, Cairo. Belonging to the Late Period of ancient Egypt, i.e. 712–323 BC, these fragments, although partially damaged, contained enough material to permit a detailed multimethod investigation that included optical microscopy, UV-induced fluorescence, multispectral imaging, SEM-EDX, XRD, and FTIR and Raman spectroscopy. The investigated fragments were decorated with blue, red, black, white, and yellow pigments, common for the art of that time. SEM-EDX allowed the identification of Egyptian blue (which was altered to greenish blue in some places), red hematite, yellow orpiment, and calcite, while FTIR spectroscopy suggested the use of animal glue as the binding medium [78].

Parchment is a writing material made from specially prepared untanned skins of animals—primarily sheep, calves, and goats. It has been used as a writing medium from the Hellenistic period to the present. In medieval Europe, it was used until the preparation and utilization of paper arrived from the Middle East, i.e. in the 13th century. As a result, a wealth of manuscripts written on parchment is considered to be one of the most precious heritages of top libraries and museums. This is the case for an illuminated parchment, a *privilegio rodado* (royal solemn privilege) preserved at the Archive of the Royal Chancellery in Granada, Spain. A complementary study that used SEM-XRF and XRD in conjunction with micro FTR spectroscopy and capillary electrophoresis provided evidence of the presence of silver and gold as well as the use of vermillion and lead minium, lead–tin yellow, azurite, and malachite. The study also found evidence for the presence of a calcite primer applied to the entire sheet and replaced by gypsum in the text area, as well as the use of iron-based ink [79].

Parchment was also used in the 16th century by the Royal Court of Portugal during the reign of Dom Manuel I of Portugal when a *foral*—a royal document valid in Portugal and its empire and used to regulate its administration, borders, and privileges—was issued. In this regard, the 1515 *foral* sent to the Portuguese village of Setubal was investigated in detail *in situ* using EDXRF and FTR micro-spectroscopy. EDXRF and FTR micro-spectroscopy evidenced the use of the relatively cheaper lead minium for red coloration and copper sulfate for green details, instead of the usual vermillion and malachite currently found on Manueline *foral* charters [80].

Two well-preserved codices from the Carolingian period (9th century) kept in the library of Kremsmünster Abbey, Upper Austria were investigated using XRF and FTR spectroscopy. Given their age and origin, such manuscripts are rarely studied. In this case, the noninvasive XRF evidenced the presence of calcium carbonates on the parchment surface, an indicator for the whitening of the animal skin, whereas the combination of XRF and Raman spectroscopy evidenced the use of iron gall ink for the black/brown text. Lead minium red ink was used for initials. Other pigments identified were lead white, hematite, azurite, and indigo, together with the rare atacamite. It worth mentioning that the presence of powders made from gold–copper and silver–copper alloys was evidenced in both gospels [81].

Gutenberg's mid-15th century invention of the movable-type printing press introduced printing to Europe, and from there, is spread all over the world. In the earliest stages of printing, i.e. before 1500, in addition to books, pamphlets, and broadsheets, illuminated incunabula appeared as a transition from illuminated manuscripts to books. Printed on paper, fine parchment, or vellum, this category of objects was closer to illuminated manuscripts, except that the text was printed and not handwritten. Given the age and rarity of such artifacts, their investigation using actual analytical methods could furnish a wealth of information concerning the materials and techniques utilized for their creation. This is the case for a 15th century illuminated incunabulum printed by the Parisian workshop of Germain Hardouyn. Among other analytical techniques used, SEM-EDX determined the composition of the gilded surfaces as well as of a multitude of pigments, such as lead white for white fields, massicot and lead–tin yellow for pale and vivid yellow details, respectively, vermillion for red details, red lead for orange spaces, a mixture of malachite, lead–tin yellow, and massicot or malachite alone for green details, azurite for blue nuances and azurite together with indigo for dark blue spaces, while goethite and carbon black were evidenced for brown and black spaces, respectively. The gilding consisted of an alloy of 95%–96% pure gold with Ag at 2%–3% and Cu in proportions varying between 1% and 2%, i.e. only high-quality pigments were found [82].

Chromolithography was one of the most successful of several methods of color printing developed during the 19th century. A detailed XRF study of such printed objects revealed the use of a relatively small number of inorganic pigments; the majority of the lithographic objects showed the presence carbon black, chrome yellow, Prussian blue, red ochre, red lead, and vermillion. At the same time, FTR spectroscopy evidenced the presence of the organic pigments indigo blue, gamboge, and a red organic pigment, pointing to the presence of an original and unusual mixture of organic and inorganic pigments [83].

Postage stamps were invented in 1840 and introduced into postal circulation in the same year. Designed to speed up the delivery of letters and parcels, postage stamps, whose numbers to date have reached billions, have become also objects well appreciated by collectors and objects of forgery. Stamps collecting is not only the hobby of millions of collectors worldwide, but also a branch of modern history, as almost all major political events are reflected by stamps. Given the extremely high value of some stamps to collectors and the possibility of forgery, XRF represents one of the most appropriate methods of investigation. XRF and SEM-EDX were useful in establishing the pigments used for the first US stamp. This permitted the identification of litharge as an orange pigment which slowly turned into black lead sulfide, as well as the identification of Al and Si from the kaolin used as a paper additive. In another example, lead minium was used for the 1d rose-red Queen Victoria stamp, a great rarity today. The same gradual transformation of red lead into lead sulfide was also evidenced by EDXRF and FTR spectroscopy in the case of the Kingdom of Prussia's first stamp [84]. As postage stamps are fragile, small objects of a great diversity of colors and origins, the best results in identifying ink pigments have been obtained through the use of polycapillary collimators [85].

Wallpaper is a type of interior decoration material that is printed or rarely hand painted and used to decorate the interior walls of domestic and public places. This custom had appeared by the end of the 17th century as a result of the Industrial Revolution and the development of the middle class. As a type of printed matter, wallpapers benefited from the advantages offered by printing techniques and the invention of new pigments. As in the cases of other manuscripts or printed matter, EDXRF and FTR spectroscopy furnished trustworthy results for the pigments and paints used as well as the wallpaper substrate. In the case of 19th century wallpapers, EDXRF evidenced the use of $BaSO_4$ as a bright white layer or background on which the pictorial layer was placed. Diffuse reflectance infrared Fourier transform spectroscopy was used to analyze blue areas; the results suggested the use of Prussian blue, a finding confirmed by EDXRF, which evidenced the presence of Fe. In addition, EDXRF confirmed the use of vermillion, attributed by Raman analysis to the rose area assigned to a mixture of vermillion and calcite compounds. Further EDXRF and Raman analysis of the brown area proved the presence of a combination of calcite, goethite, bone black, and minium, as S, Ca, Fe, and Pb, most probably obtained from ochre, chalk, and lead sulfide, were present [86].

Watercolor painting is one of the oldest painting techniques, dating from the Paleolithic. It has been used for manuscript illustration from the European Middle Ages up to the present. In contrast to the use of tempera, oil, or encaustic, watercolor uses the pigments finely dispersed in water, which can be applied to any type of media, although the most representative paintings have paper substrates. EDXRF has shown its utility in identifying the pigments used in historic watercolor paintings on paper. This was the case for Ferdinand Bauer's printed and hand-painted *Flora Graeca*, which dates from the end of 18th century [87]. As in many other cases, EDFRX determinations were accompanied by FTR spectroscopy and multispectral analysis. After an analysis of 125 paintings, these techniques applied together revealed the use of, e.g. indigo, vermillion, copper greens, copper blues, Prussian blue, red lakes, vermillion, madder, yellow lake, gamboge or ochre, and lead white [88].

The same combination of EDXRF, FTIR, and FTR spectroscopic methods was used to analyze nine different parts of the Ataturk portrait by contemporary Turkish painter Feyhaman Duran, revealing the presence of different pigments. For comparison, five paint samples collected from the artist's paint box were analyzed using the same technique. The EDXRF results for the black–green color used in portrait painting showed the presence of Ca, P, Pb, Ti, Si, Cr, S, Mn, Mg, Zn, Al, Fe, and Ba elements. The FTIR results showed the presence of chromium oxide, viridian green, manganese oxide, ivory black or bone black, chrome yellow, chalk, lead white, barite, gypsum, titanium white, clay, and zinc white. The EDXRF results for the skin color used in portrait painting showed the presence of Ti, Zn, Ca, Mg, Fe, S, Pb, Cr, Al, Si, Ba, Sr, and Cd elements. The FTIR results for the same color showed the presence of titanium white, zinc white, hematite, strontium yellow, cadmium yellow, clay, chalk, lead white, barite, and titanium white. Chemical analyses of pigment samples from the artist's paint box showed that he used mixed pigments rather than pure ones. A light yellow color was a mixture of chrome yellow and chalk; the use of EDXRF on this sample detected Ca, Pb, and Cr, which were authenticated by FTIR and Raman analyses. A deep yellow color was a mixture of orpiment, chalk, and lead white; EDXRF identified As, S, Pb, and Ca elements in this sample, which were authenticated by FTIR and Raman analyses [89].

Some painters have reused the canvases of their older paintings as support materials for newer ones. Evidencing such use is very difficult, if not almost impossible, as the new paint layers completely cover the older ones. Radiography and XRF mapping can only reveal the older layers if the pigments of the older painting contain elements that have atomic numbers as high as those of the newer ones. This was the case for Vincent Van Gogh's *Patch of Grass* (Kröller-Müller Museum, Otterlo, The Netherlands). Using synchrotron radiation and XRF, it was possible to scan the painting layer point by point, thus evidencing the 2D distribution of Sb ($Z = 51$) and Hg ($Z = 80$). The characteristic x-ray energies of these elements were high enough for the x-rays to pass through the more recent paint layers, thus depicting the underlying layers. Using a combined analysis of Zn, Sb, and Hg characteristic radiation intensities, it was possible to reconstruct the portrait of a woman with adequate fidelity. The portrait was very similar to two other portraits with similar subjects found in the collections of two other museums in the Netherlands [90].

4.4 Concluding remarks

XRF analysis in its wave-dispersive and energy-dispersive variants has shown its utility in identifying the elements that compose almost all inorganic pigments used in works of art, regardless of the x-ray generator used, i.e. isotopic soft x-ray sources or synchrotron quasi-monochromatic radiation. This kind of study has been extended to bronze and other metallic objects as well. This statement is supported by an impressive volume of investigations covering almost all types of artwork. The only relative disadvantage of XRF is its inability to determine the complete chemical composition of pigments, because the lighter elements from lithium to fluorine are very difficult to detect by XRF. This disadvantage can easily be overcome by the simultaneous use of FTIR and Raman spectroscopy; both methods are useful for the unequivocal identification of the different anions and moieties used in paint compositions. Moreover, as a result of the development of high-energy, high-resolution SDDs cooled by the Peltier effect, handheld portable XRF spectrometers permit the contactless examination of a great diversity of human artifacts, making this type of spectrometer a useful tool with which to acquire expertise in artworks. It should be noted that FTIR and Raman spectroscopic investigations can both be performed contactlessly using miniaturized instruments, which when combined with XRF can categorize almost any work of art.

References

[1] Pospisilova E, Novotny K and Porizka P *et al* 2018 Depth-resolved analysis of historical painting model samples by means of laser-induced breakdown spectroscopy and handheld X-ray fluorescence *Spectrochim. Acta* B **147** 100–8

[2] Legrand S, Vanmeert F and Van der Snickt G *et al* 2014 Examination of historical paintings by state-of-the-art hyperspectral imaging methods: from scanning infra-red spectroscopy to computed X-ray laminography *Herit. Sci.* **2** 1

[3] Li Z, Wang L and Ma Q *et al* 2014 A scientific study of the pigments in the wall paintings at Jokhang Monastery in Lhasa, Tibet, China *Herit. Sci.* **2** 1–6

[4] Aceto M 2021 Pigments–the palette of organic colorants in wall paintings *Archaeol. Anthropol. Sci.* **13** 1–23

[5] Van Elslande E, Lecomte S and Le Ho A-S 2008 Micro-Raman spectroscopy (MRS) and surface-enhanced Raman scattering (SERS) on organic colourants in archaeological pigments *J. Raman Spectrosc.* **39** 1001–6

[6] Siddall R 2018 Mineral pigments in archaeology: their analysis and the range of available materials *Minerals* **8** 201

[7] Aceto M, Agostino A and Fenoglio G *et al* 2014 Characterisation of colourants on illuminated manuscripts by portable fibre optic UV–visible–NIR reflectance spectrophotometry *Anal. Methods* **6** 1488–500

[8] Pfaff G 2017 *Inorganic Pigments* (Berlin: De Gruyter)

[9] Fontana D, Alberghina M F and Barraco R *et al* 2014 Historical pigments characterization by quantitative X-ray fluorescence *J. Cult. Herit.* **15** 266–74

[10] Donais M K and George D B 2018 *X-Ray Fluorescence Spectrometry and Its Applications to Archaeology: An Illustrated Guide* (New York: Momentum Press)

[11] Marinescu M, Emandi A and Duliu O G *et al* 2014 FT-IR, EPR and SEM–EDAX investigation of some accelerated aged painting binders *Vibrat. Spectr.* **73** 37–44

[12] Genestar C and Pons C 2005 Earth pigments in painting: characterisation and differentiation by means of FTIR spectroscopy and SEM-EDS microanalysis *Anal. Bioanal. Chem.* **382** 269–74

[13] Stuart B H 2007 *Analytical Techniques in Materials Conservation* (Chichester: Wiley)
[14] Scott D A 2016 A review of ancient Egyptian pigments and cosmetics *Stud. Conserv.* **61** 185–202
[15] Guerra M, Manso M and Pessanha S *et al* 2013 X-ray Fluorescence Spectrometry as a diagnostic tool in conservation of illuminated manuscripts *Cultural Heritage: Protection, Developments and International Perspectives* ed P Frediani *et al* (New York: Nova Science Publishers), 9 235–56 https://www.researchgate.net/publication/235917544_XRF_spectrometry_as_a_diagnostic_tool_in_conservation_of_illuminated_manuscripts (accessed 02 May 2023)
[16] Garate D 2014 Altamira and Paleolithic Cave Art of Northern Spain ed C Smith *Encyclopedia of Global Archaeology* (New York: Springer)
[17] Sauvet G, Sainz C G, Sanchidrián J L and Villaverde V 2014 Europe: Prehistoric Rock Art ed C Smith *Encyclopedia of Global Archaeology* (New York: Springer)
[18] Coulson D and Campbell A 2022 Rock Art of the Tassili 'n Ajjer, Algeria, https://rockartscandinavia.com/images/articles/coulsona10.pdf (accessed 15 September 2022)
[19] Nyberg J, MacGregor D and Azaiez F 2021 Complementary methods: γ-beam techniques, X-ray fluorescence (XRF) and nuclear magnetic resonance (NMR) *Nuclear Physics for Cultural Heritage* ed A Mackova, D MacGregor, F Azaiez, J Nyberg and E Piasetzky (Les Ulis: EDP Sciences) pp 37 53
[20] Bertin E P 1975 *Principles and Practice of X-Ray Spectrometric Analysis* (Berlin: Springer)
[21] Serafima Sister, Duliu O G and Manea M-M *et al* 2019 An XRF, XRD, FTIR, FT Raman, digital radiography and UV photography study of some classical pigments, primers and binders used in panel painting *Rom. Rep. Phys.* **71** 201 https://rrp.nipne.ro/2019/AN71201.pdf (accessed 1 April 2023)
[22] De La Calle I, Cabaleiro N and Romero V *et al* 2013 Sample pretreatment strategies for total reflection X-ray fluorescence analysis: a tutorial review *Spectrochim. Acta* B **90** 23–54
[23] Pozzi F, Rizzo A and Basso E *et al* 2021 Portable spectroscopy for cultural heritage: applications and practical challenges *Portable Spectroscopy and Spectrometry: Technologies and Instrumentation* ed R A Crocombe, P E Leary and B W Kammrath (Chichester: Wiley) pp 499–522
[24] Romano F P, Caliri C and Cosentino L *et al* 2014 Macro and micro full field X-ray fluorescence with an X-ray pinhole camera presenting high energy and high spatial resolution *Anal. Chem.* **86** 10892–9
[25] Pollard A M, Batt C M and Stern B *et al* 2007 *Analytical Chemistry in Archaeology* (Cambridge: Cambridge University Press)
[26] https://chess.cornell.edu/users/calculators/characteristic-emission-lines-elements, (accessed 14 September 2022)
[27] https://engineering.temple.edu/sites/engineering/files/Oxford%20EDS-%20Silicon%20Drift%20Detectors%20Explained.pdf (accessed 10 September 2022)
[28] Nodari L and Ricciardi P 2019 Non-invasive identification of paint binders in illuminated manuscripts by ER-FTIR spectroscopy: a systematic study of the influence of different pigments on the binders' characteristic spectral features *Herit. Sci.* **7** 7
[29] Ross C A, MacLachlan D G and Smith B J E *et al* 2020 A miniature fibre-optic Raman probe fabricated by ultrafast laser-assisted etching *Micromachines* **11** 185
[30] Serafima Sister, Duliu O G and Manea M-M *et al* 2019 Complex investigation of the five 19th century Russian-Lipovan icons *Microchem. J.* **150** 104126

[31] Rosi F, Burnstock A and Jan Van den Berg K *et al* 2008 A non-invasive XRF study supported by multivariate statistical analysis and reflectance FTIR to assess the composition of modern painting materials *Spectrochim Acta* A **71** 1655–72
[32] Bonaduce I and Andreotti A 2009 Py-GC/MS of organic paint binders. Organic mass spectrometry in art and archaeology *Organic Mass Spectrometry in Art and Archaeology* ed M P Colombini and F Modugno (Chichester: Wiley) pp 303–26
[33] Mahnke H E, Denker A and Salomon J 2009 Accelerators and X-rays in cultural heritage investigations *C. R. Phys.* **10** 660–75
[34] Hiley C I, Hansford G and Estaugh N 2021 High-resolution non-invasive X-ray diffraction analysis of artists' paints *J. Cult. Her* **53** 1–13
[35] Gasanova S and Hermon S 2017 Micro-XRF complemented by X-radiography and digital microscopy imaging for the study of hidden paintings *Proc. SPIE* **10331** 103310T
[36] Anitha A, Brasoveanu A and Duart M *et al* 2013 Restoration of X-ray fluorescence images of hidden paintings *Sign. Proces* **93** 592–604
[37] Lizun D 2021 From Paris and Shanghai to Singapore: a multidisciplinary study in evaluating the provenance and dating of two of Liu Kang's paintings *J. Conserv. Sci.* **37** 322–39
[38] Delaney J K, Dooley K A and Radpour R *et al* 2017 Macroscale multimodal imaging reveals ancient painting production technology and the vogue in Greco-Roman Egypt *Sci. Rep.* **7** 15509
[39] Pitzalis D, Cassan I and Thomas L *et al* 2008 Non-destructive analyses for modern paintings: the Russian Avant-garde case *9th Int. Conf. on NDT of Art 2008 Jerusalem (Israel, May 2008)* (accessed 21 September 2022) https://www.ndt.net/search/docs.php3?id=6094
[40] Longoni M, Beccaria C and Bonizzoni L *et al* 2022 Silver monochrome 'Concetto spaziale' by Lucio Fontana: a spectroscopic non-and micro-invasive investigation of materials *Molecules* **27** 4442
[41] Bernady E, Goryl M and Walczak M 2021 XRF imaging (MA-XRF) as a valuable method in the analysisof nonhomogeneous structures of grisaille paint layers *Heritage* **4** 3193–207
[42] Olivares M, Castro K and Corchon M S *et al* 2013 Non-invasive portable instrumentation to study Palaeolithic rock paintings: the case of La Peña Cave in San Roman de Candamo (Asturias, Spain) *J. Archaeol. Sci.* **40** 1354–60
[43] Serafima Sister, Duliu O G, Manea M M and Niculescu G 2016 FTIR, XRF and optical microscopy analysis of the painting layer of an early 19th century icon *Rom. Rep. Phys.* **68** 191–202 https://rrp.nipne.ro/2016_68_1/A15.pdf (accessed 21 September 2022)
[44] Al-Emam E, Beltran V and De Meyer S *et al* 2021 Removal of a past varnish treatment from a 19th-century Belgian wall painting by means of a solvent-loaded double network hydrogel *Polymers* **13** 2651
[45] Mauran G, Lebon M and Détroit F *et al* 2019 First *in situ* pXRF analyses of rock paintings in Erongo, Namibia: results, current limits, and prospects *Archaeol. Anthropol. Sci.* **11** 4123–45
[46] Stuart B H and Thomas P S 2017 Pigment characterization in Australian rock art: a review of modern instrumental methods of analysis (review) *Herit. Sci.* **5** 10
[47] Alonso A, Pérez N A and Sil J L R *et al* 2014 Comparative spectroscopic analysis of Maya wall paintings from Ek'Balam, Mexico *Mater. Res. Soc. Symp. Proc.* **1618** 63–72
[48] Cameron M A S, Jones R E and Philippakis S E 1977 Scientific analyses of Minoan fresco samples from Knossos *Annu. Br. Sch. Athens* **72** 121–84

[49] Cortea I M, Ratoiu L and Ghervase L *et al* 2021 Investigation of ancient wall painting fragments discovered in the Roman Baths from Alburnus Maior by complementary non-destructive techniques *Appl. Sci.* **11** 10049
[50] Siddal R 2018 Mineral pigments in archaeology: their analysis and the range of available materials *Minerals* **8** 201
[51] Veneranda M, Costantini I and Prieto-Taboada N *et al* 2021 Spectroscopic-assisted archaeometric studies to determine the production technology of the VI BC Zeus Enthroned statue (Paestum, Italy) and Pre-Roman technology transfer *Spectrochim. Acta* A **249** 119294
[52] Nuevo M J, Sanchez A M and Ojeda M A *et al* 2016 Spectroscopic analysis of decorated vestiges found in the Roman Theatre of Medellin, Badajoz, Spain *Microchem. J.* **124** 675–81
[53] Cechák T, Trojek T and Ružek V *et al* 2022 X-ray fluorescence analysis of pigments in gothic frescoes, coats of arms, and polychromy on sculptures on the Triforium in St. Vitus Cathedral on the territory of the Czech Kingdom in the 14th century *Materials* **15** 5033
[54] Costantini I, Aramendia J and Tomasini E *et al* 2021 Detection of unexpected copper sulfate decay compounds on late Gothic mural paintings: assessing the threat of environmental impact *Microchem. J.* **169** 106542
[55] Cennini C 2001 *Il libro dell'arte* ed F Brunello (Milan: Neri Pozza) p 141 https://neripozza.it/libri/il-libro-dellarte
[56] Cesareo R, Castellano A and Buccolieri G *et al* |2004 Portable equipment for energy dispersive X-ray fluorescence analysis of Giotto's frescoes in the Chapel of the Scrovegni *Nucl. Instr. Meth. Phys. Res.B* **213** 703–6
[57] Khramchenkova R, Ionescu C and Sitdikov A *et al* 2019 A pXRF *in situ* study of 16th–17th century fresco paints from Sviyazhsk (Tatarstan Republic, Russian Federation) *Minerals* **9** 114
[58] Barbu O H 2021 pXRF and FTIR spectrometry applied to the study of azurite and smalt in Romanian medieval wall painting *Conservation and Painting Techniques of Wall Paintings on the Ancient Silk Road, Cultural Heritage Science* ed S Aoki *et al* (Berlin: Springer) pp 105–17
[59] Christopoulou E, Ganetsos T and Nikolaos Laskaris N Non-destructive XRF and Raman spectroscopy analysis in pigment Identification of a wall painting of the painter Nikiforos Lytras from the Chapel of Agios Georgios, Haidari, Athens *Archaeology* **9** 7–13
[60] Vandiver P B, Soffer O and Klima B *et al* 1989 The origins of ceramic technology at Dolni Dolni Vestonice, Czechoslovakia, Czechoslovakia *Science* **246** 1002–8
[61] Angeli L, Brunetti A and Legnaioli S *et al* 2019 Analysis of the middle Neolithic trichrome pottery: characterization of the decoration using X-ray fluorescence and Raman spectroscopy *J. Archaeol. Sci. Rep.* **24** 192–7
[62] Stos-Fertner Z, Hedges R E M and Evely R D G 1979 The application of the XRF-XRD method to the analysis of the pigments of Minoan painted pottery *Archaeometry* **21** 187–94
[63] Shoval S and Gilboa A 2016 PXRF analysis of pigments in decorations on ceramics in the East Mediterranean: a test-case on Cypro-Geometric and Cypro-Archaic Bichrome ceramics at Tel Dor, Israel *J. Archaeol. Sci. Rep.* **7** 472–9
[64] Pincé P, Vekemans B and Vandenabeele P *et al* 2016 Analysis of pre-Islamic ceramics from the Kur River Basin (Fars, Iran) using handheld X-ray fluorescence spectrometry *Spectrochim. Acta* B **123** 150–6
[65] Vázquez de Ágredos-Pascual M L, Roldán-García C and Murcia-Mascarós S *et al* 2019 Multianalytical characterization of pigments from funerary artefacts belongs to the

Chupicuaro Culture (Western Mexico): oldest Maya blue and cinnabar identified in pre-Columbian Mesoamerica *Microchem. J.* **150** 104101
[66] Morón A G, Polvorinos del Río A J and Castaing J *et al* 2016 Ceramics by Niculoso Pisano and quantitative analysis of glazes using portable XRF *PH Investig.* **6** 1–23 http://iaph.es/revistaph/index.php/revistaph/issue/view/106
[67] Simsek G 2020 EXPRESS: handheld X-ray fluorescence (XRF) versus wavelength dispersive XRF: characterization of Chinese blue and white porcelain sherds using handheld and Laboratory-type XRF instruments *Appl. Spectr.* **74** 314–22
[68] Yap C T and Tang S M 1984 X-Ray fluorescence analysis of modern and recent Chinese porcelains *Archaeometry* **26** 78–81
[69] Aggelakopoulou E, Sotiropoulou S and Karagiannis G 2022 Architectural polychromy on the Athenian Acropolis: an *in situ* non-invasive analytical investigation of the colour remains *Heritage* **5** 756–87
[70] Fittschen U, Scharf O and Menzel M 2015 Full-field X-ray fluorescence microscopy using a color X-ray camera *Microsc. Today* **23** 36–43
[71] Reiche I, Müller K and Albéric M *et al* 2013 Discovering vanished paints and naturally formed gold nanoparticles on 2800 years old Phoenician ivories using SR-FF-MicroXRF with the color X-ray camera *Anal. Chem.* **2013** 5857–66
[72] Calza C, Oliveira D F and Freitas R P *et al* Analysis of sculptures using XRF and X-ray radiography *Rad. Phys. Chem.* **116** 326–31
[73] Križnar A, Muñoz M V and de la Paz F *et al* 2009 XRF analysis of two terracotta polychrome sculptures by Pietro Torrigiano *X-Ray Spectr.* **38** 169–74
[74] General-Toro P, Bordalo R and Moreira P R *et al* 2021 Art casting in Portuguese 19th century industrial foundries: a multi-analytical study of an emblematic copper-based alloy monument *Heritage* **2021** 3050–64
[75] Heginbotham A, Erdmann R and Hayek L-A C 2018 The dating of French gilt bronzes with ED-XRF analysis and machine learning *J. Am. Inst. Conserv.* **57** 149–68
[76] Nakamura R and Naruse M 2018 Spectroscopic analysis of colorants used for bachiru carving technique found in the Shosoin treasures *Stud. Conserv.* **63** 267–76
[77] Schiavon N, Panganiban P and Valadas S *et al* 2021 A multi-analytical study of Egyptian funerary artifacts from three Portuguese museum collections *Heritage* **4** 2973–95
[78] Ali M F, Darwish S S and El Sheikha A M 2020 Multispectral analysis and investigation of overlapping layer cartonnage fragments from Egyptian museum, Cairo *Sci. Cult.* **6** 25–36
[79] Duran A, López-Montes A and Castaing *et al* 2014 Analysis of a royal 15th century illuminated parchment using a portable XRF–XRD system and micro-invasive techniques *J. Archaeol. Sci.* **45** 52–8
[80] Guerra M, Carvalho M L and Le Gac A *et al* 2016 New insights into the red and green pigments in the illuminated foral charter of Setubal (1515) by combined use of μ-Raman and X-ray fluorescence spectrometry *Appl. Phys.* A **119** 104902
[81] Fruhmann B, Cappa F and Vetter W *et al* 2018 Multianalytical approach for the analysis of the Codices Millenarius Maior and Millenarius Minor in Kremsmuenster Abbey, Upper Austria *Herit. Sci.* **6** 10
[82] Miguel C, Bottura S and Ferreira T *et al* 2019 Unveiling the underprintings of a late-fifteenth-early-sixteenth century illuminated French incunabulum by infrared reflectography *J. Cult. Herit.* **40** 34–42

[83] Pitarch A, Alvarez-Perez A and Castro K *et al* 2012 Raman analysis assessed by Fourier-transformed infrared and X-ray fluorescence spectroscopies: a multi-analytical approach of ancient chromolithographs from the 19th century *J. Raman Spectrosc.* **43** 411–8
[84] Groß U 2018 Forensic philately: the science of stamp authentication *Chem. Texts* **4** 18
[85] Cesareo R and Brunetti A 2008 X-ray fluorescence—analysis of 19th century stamps *X-Ray Spectrom* **37** 260–4
[86] Arrizabalaga I, Gomez-Laserna O and Aramendia J *et al* 2014 Determination of the pigments present in a wallpaper of the middle nineteenth century: the combination of mid-diffuse reflectance and far infrared spectroscopies *Spectrochim. Acta* A **124** 308–14
[87] Sibthorp J and Buer F 1806 Flora Græca: sive Plantarum rariorum historia, quas in provinciis aut insulis Græciæ, Typis Richardi Taylor et socii, in Vico Shoe-Lane : Veneunt apud Johannem White, in Vico Fleet-Street, Londini https://archive.org/details/floragraeca-sive1sibt (accessed 16 October 2022)
[88] Mulholland R, Howel D and Beeby A *et al* 2017 Identifying eighteenth century pigments at the Bodleian library using *in situ* Raman spectroscopy, XRF and hyperspectral imaging *Herit. Sci.* **5** 43
[89] Akyuz S, Akyuz T and Emre G *et al* 2012 Pigment analyses of a portrait and paint box of Turkish artist Feyhaman Duran (1886–1970): the EDXRF, FT-IR and micro Raman spectroscopic studies *Spectrochim. Acta* A **89** 74–81
[90] Dik J, Janssens K and Van Der Snickt G *et al* 2008 Visualization of a lost painting by Vincent van Gogh using synchrotron radiation based X-ray fluorescence elemental mapping *Anal. Chem.* **80** 6436–42
[91] Janssens K, Van der Snickt G and Vanmeert F *et al* 2016 Non-invasive and non-destructive examination of artistic pigments, paints, and paintings by means of X-ray methods *Top. Curr. Chem. (Z)* **374** 81

IOP Publishing

Spectroscopic and Microscopy Techniques for Archaeological and Cultural Heritage Research (Second Edition)

Ashutosh Kumar Shukla

Chapter 5

ESR dating in Archaeology

Verónica Guilarte

Electron spin resonance (ESR) dating is a chronometric or paleodosimetric dating method, based on the evaluation of the exposure of some materials to natural ionizing radiation. In ESR dating, the damaged produced by the radiation in the crystal lattice of a material is quantified, by measuring the number of paramagnetic centers accumulated over time. Since the first study conducted by Ikeya (1975) regarding dating several stalactites, numerous applications have been developed using a broad range of materials, such as quartz, mollusc shells, corals, speleothems and fossil tooth enamel. In addition, ESR dating shows a great potential for archaeology since it could cover the last 2.6 Ma in any geological context. In this chapter, the principles of ESR dating, the time frame, and the advantages and drawbacks of ESR dating in comparison with other absolute dating methods will be analyzed. Moreover, the main applications of ESR dating in archaeology, the future scope and the potential of this interesting absolute dating method will be presented.

Definitions

D_E	Equivalent dose. The total absorbed radiation dose as measured in the laboratory
Ka	1000 years
Ma	1 000 000 years

5.1 Introduction

Geochronology has become a key discipline in archaelogy and therefore, numerous chronometric methods are available nowadays for dating archaeological and geological materials. It is not possible to describe a '*perfect dating method*' for all situations, since all of them show advantages and drawbacks. Radiocarbon (C-14),

doi:10.1088/978-0-7503-4846-1ch5

U-series (U–Th) and argon–argon (Ar–Ar) could probably provide the more accurate results, however they cannot be used in all geological context and with all types of materials. In addition, the chronology of the site plays an important role in order to decide the more appropriate dating method. It is in this context that other important chronometric methods have been developed, such as luminescence (optical stimulated luminescence: OSL and thermoluminesce: TL), terrestrial cosmogenic nuclides or electron spin resonance (ESR).

Electron spin resonance (ESR) dating is an absolute dating method. Different authors have classified it as a radiation exposure dating method, a trapped charge dating method or a paleodosimetric method. Similar to the methods using luminescence phenomena, ESR dating is based on the evaluation of the exposure of some materials to natural radioactivity. This exposure is quantified in terms of absorbed radiation dose, which corresponds to the energy deposited in a medium by ionizing radiations. In ESR dating, the sample acts as a dosimeter, i.e. a material able to register the energy absorbed when exposed to natural radioactivity. Indeed, under the effect of natural radioactivity, some materials acquire or develop a paramagnetic behavior, which may be detected and quantified by ESR spectroscopy.

The use of ESR spectroscopy (also known as electron paramagnetic resonance: EPR) for dating purposes was first suggested by Ikeya (1975), who published a pioneering work based on the study of a few stalactites from Japanese caves. After this moment, numerous applications to a wide range of materials, such as quartz, flints, carbonate crystal and fossil remains, have been investigated (see extensive reviews in Grün 1989; Ikeya 1993; Rink 1997), and nowadays ESR dating has become a landmark in the study of *Quaternary Geochronology*. ESR dating shows a great potential for archaeology since it could cover the last 2.6 Ma in any geological context. The objective of this chapter is to present the principles of ESR dating, and an updated overview of the potential and current limitations of the ESR dating method used in geoarchaeological contexts, through a review of different studies carried out over the last decades on different archaeological sites. A special focus will be given to ESR applications using fossil teeth and optically bleached quartz grains, because they have become the most useful materials for ESR dating in the whole Quaternary.

5.2 Principles of ESR dating

In this section, the basis of ESR dating is described, including the principles of ESR spectroscopy with a dating purpose. In addition, the whole general protocol, from the sampling to the age calculation, is presented as well as the main parameters required for the age assessment.

5.2.1 Basic principles

ESR spectroscopy is a technique based on the detection of systems with unpaired electrons (i.e. paramagnetic systems). To measure the unpaired electrons, the sample should be placed in the magnetic field and exposed to microwaves of a fixed frequency. At a given magnetic field value, corresponding with the resonance condition ($\Delta E = h\upsilon = g\mu_B B_0$, where g is the g-factor, μ_B is the Bohr magneton and B_0 is the magnetic field), there is an absorption of part of the microwave radiation by the sample. The radiation changes after interacting with the sample give information about the sample. In ESR, the absorption of radiation by systems with unpaired electrons is studied; and the intensity of the ESR signal is proportional to the number of unpaired electrons in the sample. Electron spin resonance is a powerful technique and is frequently used in different scientific fields: chemistry, physics, biology, medicine and material science, but its use for retrospective dosimetry, and dating archaeological materials is one of the most original applications.

In ESR dating, the paramagnetic centers can be formed in minerals by ionizing radiation (figure 5.1): a mineral, which is an insulator, shows two energy levels where electrons might occur, the valence band (ground state) and the conduction band. At the time when the mineral is formed, all electrons are in the ground state. However, the interaction of the mineral with ionizing radiation (α and β particles; and γ and cosmic rays), naturally emitted by the sample and its surrounding environment, induces changes in the electronic structure of the material, and some electrons are transferred to the conduction band. After a short time of diffusion, these electrons recombine with holes near the valence band (figure 5.1). However, all the natural minerals or crystals, such as hydroxyapatite, which is the main component of tooth enamel, contain defect sites (vacancies, impurities, interstitial atoms). It is in these defect sites where electrons and holes can be trapped (e.g. Grün 2006). The number of trapped electrons and holes is linked to the amount of radiation received by the crystal. These trapped electrons give rise to characteristic ESR signals.

The number of trapped electrons and, hence, the intensity of the ESR signal is directly proportional to: (1) the strength of the natural radiation in the sample and its surroundings (dose rate), (2) the total number of traps in the mineral or crystal

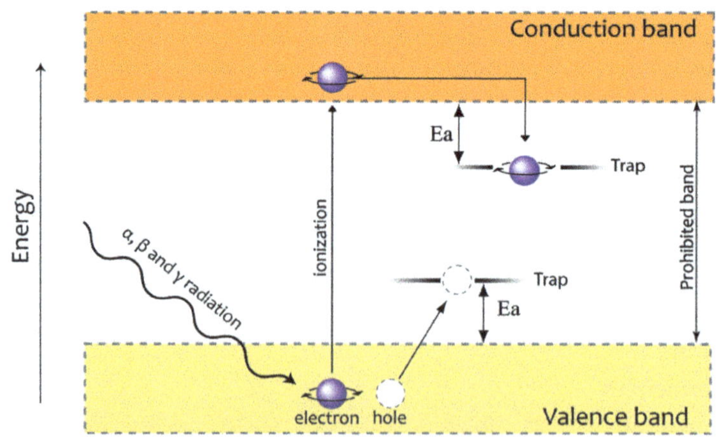

Figure 5.1. Energy band model of trapping charges process.

(sensitivity), (3) the exposition time (age of the material). An ESR age can be calculated from the following equation:

$$D_E = \int_0^T D(t)dt. \tag{5.1}$$

In equation (5.1), D_E is the equivalent dose or paleodose: expressed using the dosimetry unit: Gray or Gy (which represent the absorption of 1 J of radiation energy per kilogram of matter): the total dose that the sample has received during the time elapsed between the zeroing of ESR clock ($t = 0$) and the sampling ($t = T$). $D(t)$ is the dose rate (usually expressed in Gy/ka or μGy/a). If it is assumed that the dose rate is constant over time, equation (5.1) can be expressed as follows:

$$D_E = DT, \tag{5.2}$$

where T is the age of the sample and D the dose rate.

The D_E value is determined thanks to the ESR spectroscopy and the dose rate is calculated from the analysis of the radioactive elements (mainly Th, U and ^{40}K) in the sample and its surroundings.

Different authors describe that four different processes could lead to the elimination of paramagnetic centers and then, to the zeroing of the geochronological clock ($t = 0$): (a) crystallization, (b) optical bleaching, (c) mechanical stress or annealing and (d) action of heat (Ikeya 1993). It means that it is possible to date different events, such as a crystallization process (a), the last exposure to sunlight (b), a fault or glacier movement (c), or the last time a material was heated (d).

In ESR dating, the sample behaves as a natural dosimeter, which means that the sample is able to record the natural radiation of the sample's surrounding environment and the sample itself, from the time the system was reset (clock, $t = 0$). Being the sample considered a dosimeter, ESR dating is also known as a paleodosimetric method.

A number of different geological and archaeological materials can be studied using ESR with dosimetric and/or geochronological purposes, among them: carbonates (speleothems, travertines, calcite, foraminifera, mollusc shells, corals), phosphates (fossil teeth, apatite), silicates (quartz, burned flint, feldspars, clay minerals) and sulfates (gypsum) (see for example Grün 1989; Ikeya 1993).

A sample should have some properties in order to be dated. Grün (1989) defined six basic principles that any ESR signal must satisfy to provide reliable dose estimations:

- The initial signal (at $t = 0$) is either zero or can be experimentally calculated.
- The ESR signal intensity increases in proportion to the dose received.
- The ESR signal must have a thermal stability, that is at least one order of magnitude higher than the age of the sample.
- The number of traps is constant. Recrystallization, crystal growth or phase transitions should not have occurred.
- The ESR signal must not show anomalous fading.

- ESR signal are not affected by chemical and physical preparation (grinding, chemical reactions, laboratory light exposure, etc). However, if there is any influence, it must be controlled and quantified.

Two processes limit the applicability of ESR dating: (1) **Saturation**: when all traps are filled, any further irradiation does not result in the formation of additional paramagnetic centers. Saturation depends on the environment. Therefore, saturation may happen within a short period of time (a few thousand years) or, in contrast, only after a few million years in low dose rate environments. (2) **Thermal stability**: trapped electrons have only a limited probability to stay in the traps. Only if the *mean lifetime* is approximately 10 times larger than the age of the sample can the effect of fading be neglected.

It is not possible to give general upper age limits for ESR dating because saturation and thermal stability depend on the specific nature of the centers and environment of the sample (Grün 2007). Each case needs to be evaluated individually. ESR measurements of tooth enamel have been carried out in accident dosimetry for the assessment of small doses. This dose amount in a natural environment would be accumulated within hundred years, which may be suggested as the youngest tooth samples to be dated. ESR dating of mollusc shells and corals has provided good results for samples as young as a few hundred years. On the oldest end of the dating range, ESR analysis of tooth enamel has provided reasonable results in the 2 million years range (Curnoe *et al* 2001; Schwarcz *et al* 1994), and even ages of several million years for optically bleached quartz grains in a low dose rate environment. Corals and shells produce systematic age underestimations once the samples are older than approximately 300 ka.

5.2.2 ESR age determination

The process of an ESR age estimation implies to determine two main parameters: (1) equivalent dose (D_E) or total dose that the sample has received during its history, from $t = 0$, and (2) the dose rate (D). Although ESR gives the name to the whole dating method (ESR dating), ESR spectroscopy is only used to determine the D_E value.

5.2.2.1 $\mathrm{D_E}$ *determination*

The D_E reconstruction process using ESR spectroscopy is divided into different steps: (a) sampling, (b) physical and chemical sample preparation, (c) laboratory irradiation, (d) ESR measurements and, (e) data processing to achieve the D_E value.

(a) Sampling

Fieldwork is the first step of the dating procedure. Not only does it consist in collecting the sample, but it is also required to register all the relevant information for data analysis and interpretation, such as sedimentary context, homogeneity, texture and composition of the sediment, sample depth, etc. The sampling process could be carried out in many different ways, depending on the type of material to be analyzed (Grün 1989). For example, in the case of sampling sediments for ESR dating of quartz grains, it is important to minimize the exposure of the raw sediment to sunlight. It is known several strategies that can be employed, among them: (a) inserting a PVC tube into the section, (b) extracting an intact block of sediment and removing in the laboratory the block's part that was in contact with sunlight, all of it under strict light conditions, (c) collecting the sediment under an opaque cover. It is necessary to highlight that all the samples should be kept in opaque bags and containers during storage. For teeth dating, most of the samples have already been collected during an archaeological or palaeontological excavation and then, should be chosen from collections. Therefore, it is important that the selected tooth is well preserved and the precise site location is correctly registered in order to avoid future complications during the subsequent sample preparation and environmental dose rate reconstruction.

On the other hand, it is also advisable to carry out *in situ* measurements of the natural radioactivity at the place where the sample was collected, or the nearest possible location, to obtain an accurate value of the gamma dose rate. The gamma dose rate could be registered by:

- A portable gamma spectrometer, with NaI or $LaBr_3$ scintillation probes (figure 5.2(a)), and using one of the following techniques: (1) the *energy windows technique*, which provides information of the specific activities of each radioelement (Arnold *et al* 2012); or (2) the *threshold technique*, which gives a total gamma dose rate (Duval and Arnold 2013). For both techniques, the registration time could be even less than 1 h.
- Synthetic TL dosimeters (e.g. $CaSO_4$:Dy, Al_2O_3:C, figure 5.2(b)), and registering a total gamma dose rate over a long period (up to several month, e.g. Miallier *et al* 2009).

Moreover, it is required to collect small quantities of sediment (200–300 g) for future laboratory analysis, such as determining the water content or the radionuclides activity concentration of the sediment surrounding the sample (the latter using, for example, a high-purity germanium detector, HpGe).

Figure 5.2. Examples of techniques that are used in the field to assess the dose rate. (a) Field measurement using a portable gamma spectrometer (Canberra InSpector multichannel analyzer connected to a NaI(Tl) probe). (b) TL dosimeters ($CaSO_4$:Dy encapsulated in copper).

(b) Sample preparation

The main objective of sample preparation is to extract 'pure' material (a few grams, or at least 100 mg in the case of tooth enamel) with no contamination that would interfere with the following analysis (ESR, U-series, etc). For ESR dating, the sample preparation steps depend on the type of material to be analyzed. Grün (1989) described the sample preparation processes for different materials, showing a wide range of tools and procedures. For example, the preparation of fossil teeth, usually implies tools such as a dentist drill and a Vernier scale; and different procedures: cutting, tissue separation, removal of enamel surface, tissue thickness measurements, grounding, sieving, etc. In the case of the extraction of quartz grains from sediment, the protocols involve several steps, among them: sieving, different chemical reactions in order to remove: the carbonates (using HCl) and the organic matter (using H_2O_2), different density separations to separate quartz from heavy minerals and feldspars (using SPT: sodium polytungstate), magnetic separation to eliminate magnetic minerals, removal of the outermost layer of quartz grains (with HF) and sieving (Duval *et al* 2017). In addition, the elimination of the outer part of a sample, during preparation, allows to minimize or even eliminate the contribution of the external alpha and beta particles radiation (see Dose rate determination section).

(c) D_E **reconstruction**

An ESR spectrometer (shown in figure 5.3) is the equipment used to determine the D_E value. The main components of an ESR spectrometer are: (1) the source of the electromagnetic radiation, (2) the detector, (3) the sample placed in a resonator, (4) the electromagnet system and (5) the console. The electromagnetic radiation source and detector are in a box called the 'microwave bridge'. The sample is placed inside a microwave resonator, which is a metal box that helps to amplify weak signals from

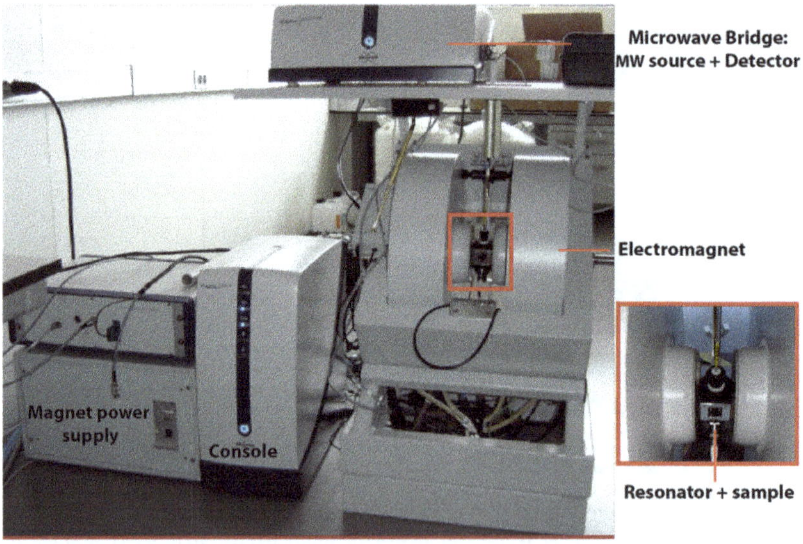

Figure 5.3. ESR spectrometer and its main components.

the sample. There is also an electromagnet that produces the magnetic field. The magnet 'tunes' the electronic energy levels, and allows to achieve the resonant condition (see section 5.2.1). In addition, the signal processing and control electronics are placed in the console.

In ESR dating the materials to be studied must behave as natural dosimeters, which means they are able to record the radiation over time. The standard procedure to determine the D_E value involves artificially 'aging' the sample by irradiating one or several subsample aliquots (single aliquot or multiple aliquot techniques) at various doses (*additive dose method*) in order to study the behaviour of the ESR signal with increasing dose values. The irradiation process generates more paramagnetic centers; consequently, the ESR signals increase. The ESR signal intensity measured is extracted from the spectrum obtained for each aliquot (figure 5.4(a)). The paired data (dose, intensity) are plotted as shown in figure 5.4(b). Then, a curve could be fitted through the experimental data points in order to describe the behaviour to the sample with the increase of doses. Further details about the fitting procedure and the corresponding functions can be found in Duval *et al* (2009). Finally, the extrapolation to zero ESR intensity allows to determine the D_E value from the intersection of the best fit of the data with the dose axis. The name D_E comes from the fact that the laboratory procedures use gamma sources, whereas the dose that the sample has received in the past is a combination of different types of radiation (alpha and beta particles and gamma and cosmic rays). Therefore, the experimentally calculated dose value is the gamma equivalent dose of the naturally received dose, which is called paleodose.

Another possible method to assess the D_E value consists in resetting the ESR signal to zero in the natural sample (the sample extracted from the site) prior to any laboratory irradiation (*regenerative dose method*). However, contrary to luminescence studies, in which this regenerative method is the most common, the heating process required for resetting the signals to zero or a known value, is often not possible. This is due to the fact that heating the sample could cause changes in the

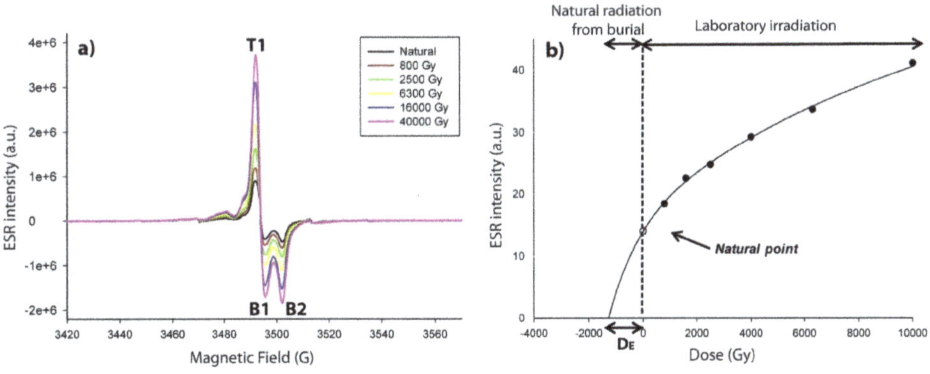

Figure 5.4. ESR dose reconstruction. (a) Evolution of the ESR spectra of tooth enamel with the irradiation dose. The ESR intensities are extracted from each ESR spectrum by measuring peak-to-peak amplitudes between T1 and B2 peaks. (b) Dose-response curve (DRC) and D_E determination process.

ESR properties of the minerals, such as the creation of new radicals or the removal of water molecules.

In ESR spectroscopy, the trapping process is limited by the number of traps available in the sample crystal. In addition, it is also necessary to consider the ability to trap more electrons, which is inversely proportional to the number of electrons already trapped. Therefore, the dose-response is not linear but is most often described by one or several saturating functions.

ESR measurements for dating purpose are traditionally performed with X-band spectrometers ($\upsilon \sim 9$–10 GHz) and using powder samples (grain size between 50 and 500 μm). The latter with the objective to minimize the angular dependence of the ESR signal with the sample position inside the resonator and then to improve measurement reproducibility.

(d) ESR measurements

In quantitative ESR, there are different parameters that should be optimized in order to achieve accurate ESR measurements and obtain the best signal/noise ratio for the sample ESR signals. These parameters can be divided into three different groups: (1) Microwave-related parameters: microwave frequency and microwave power; (2) Magnetic field parameters, such as the magnetic field value at resonance, the magnetic field sweep width and the sweep time; and (3) Signal channel parameters: for example the receiver gain, the spectrum resolution, the modulation frequency, the modulation amplitude, the number of repeated measurements, among others.

In addition, in ESR dating, it is crucial to ensure constant experimental conditions, in order to eliminate, or at least minimize, the variations of the ESR signal that are induced by the system or its environment, and to make sure that only the variations derived from the sample itself are recorded by the ESR spectrometer. The main factors that could impact measurement repeatability could be summarized as follows:

1. *Stability of the measurement temperature over time*: the temperature circulating in the magnets should be controlled and kept stable over time (usually with the help of a chiller). The temperature of the room in which measurements are performed should be also stable over time (preferable not higher than 20 °C). In addition, there are some materials with paramagnetic centers that have to be measured at low temperature, usually with liquid nitrogen and at temperature around 88–110 K, such as the aluminum and titanium paramagnetic centers in quartz grains. For these samples, the ESR intensity strongly depends on the measurement temperature (Duval and Guilarte 2012). It means that the selected measurement temperature should be controlled and kept stable during ESR measurements in order to achieve reproducible results (for example with a variable temperature unit system working with liquid nitrogen). However, even if the temperature control units are quite stable over time, a small temperature variations during measurements bigger than 0.2–0.3 K should be recorded. Then, ESR

intensity should be corrected accordingly in order to achieve reproducible results (Duval and Guilarte 2012).

5. *Humidity conditions* should be controlled, and in case of high humidity levels in the measurement room, the use of a dehumidifier is advisable. Another option is to perform the ESR measurements under a nitrogen atmosphere to avoid the water condensation inside the resonator.
3. *Vertical position of the sample within the resonator*: because the resonator has a non-uniform volume sensitivity distribution (Eaton *et al* 2010), it is essential to obtain the signal intensity distribution of the resonator along the sample tube axis (i.e. vertically) with the aim of ensuring the sample is actually positioned at the center of the resonator (figure 5.5). It implies that: (i) the sample center should match with the resonator center in order to obtain the highest ESR intensity and (ii) the sample should be systematically placed at the same position within the resonator to achieve repeatable measurements (Guilarte and Duval 2020). In addition, since the ESR intensity is proportional to the sample density in the middle of the resonator, some care should be taken to

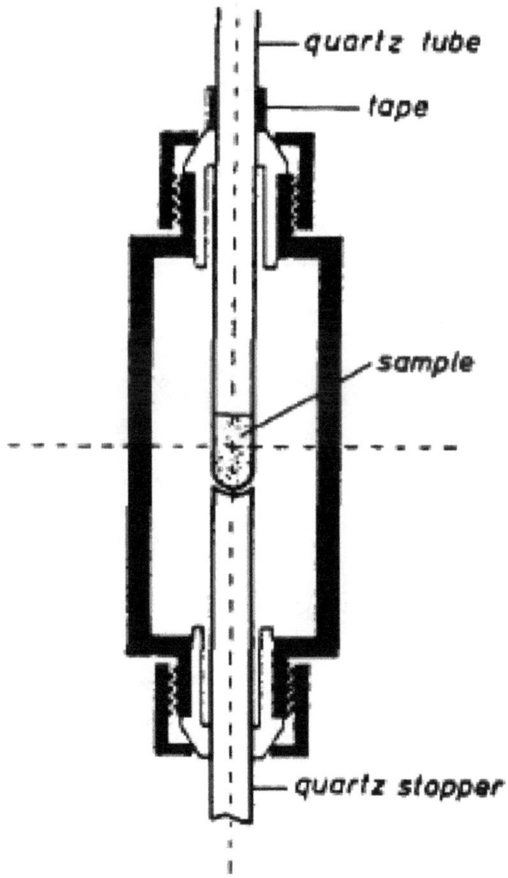

Figure 5.5. Positioning of the sample inside the resonator. Reprinted from Grün (1989), Copyright (1989), with permission from Elsevier.

pack the sample. Regulla *et al* (1985) showed that the signal intensity of a powder (<75 μm) might vary up to 100% due to different packing, and also that an angular dependence of the ESR signal (with respect to the orientation to the outer magnetic field) was present when using large grains (>1 mm). Using a sieved fraction of 100–300 μm avoids both problems.

5.2.2.2 Dose rate determination

The dose rate, D, is calculated from the concentrations of radioactive elements in the sample itself and its surrounding environment. For the calculations, only the U and Th decay chains and the ^{40}K are significant (figure 5.6).

In addition, there is a component of cosmic rays that should be taken into consideration. Different ionizing radiations are emitted from the radioactive elements: (1) alpha particles, which have a very short penetration range (20–40 μm). Alpha particles are less efficient to produce ESR intensity than beta particles and gamma rays. (2) Beta particles (electrons), which have a penetration range of around 2 mm; and (3) gamma rays, with a penetration range in matter about 30 cm.

The dose rate could be obtained as follows:

$$D = D_{\text{int}} + D_{\text{ext}} + D_{\text{cos}} = D_\alpha + D_\beta + D_\gamma + D_{\text{cos}}. \quad (5.3)$$

Where D_{int}, D_{ext} and D_{cos}: are the internal, external and cosmic dose rates; and D_α, D_β and D_γ: are the alpha-particle, beta-particle, and gamma-ray components to the total dose rate (D), respectively.

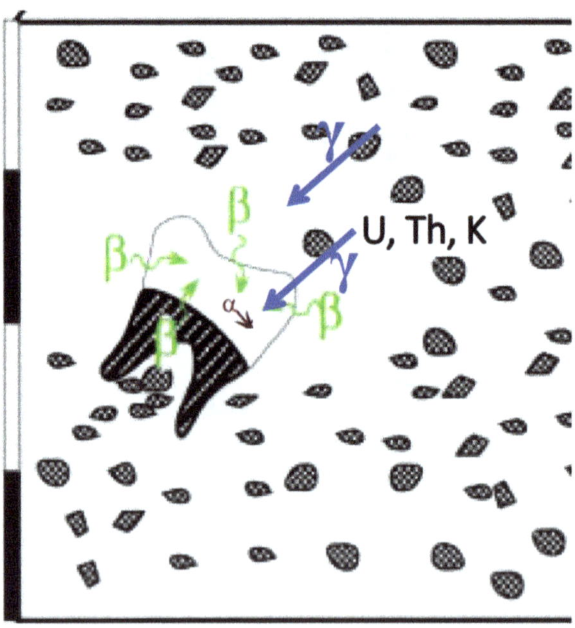

Figure 5.6. Schematic representation of the surrounding environment doses received by a fossil tooth in an archaeological site. Adapted from Grün (2007), Copyright (2007), with permission from Elsevier.

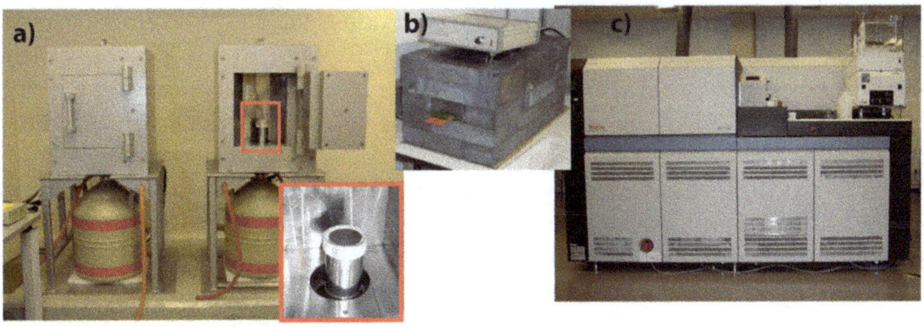

Figure 5.7. Examples of laboratory techniques for the dose rate assessment. (a) Gamma spectrometry: high-purity germanium detectors (Canberra). (b) Beta counter. (c) ThermoFinnigam Neptune multicollector (MC) inductively coupled plasma mass spectrometry (ICP-MS).

The concentrations of radioactive elements could be obtained in the laboratory using different techniques (Goldstein and Stirling 2003) (figure 5.7). For example, counting the number of particles emitted (using alpha spectrometry or gamma spectrometry) or directly from the isotopes using mass spectrometry (TIMS or ICP-MS). A beta counter also allows to obtain a total beta or gamma dose rate value (Ankjærgaard and Murray 2007).

The conversion of concentration data into dose values are calculated following the published factors (Guérin *et al* 2011 and references therein). The concentrations of radioactive elements within a sample are usually very different from the surroundings, which means that the internal and external doses must be calculated independently. Then, the external dose rate is a combination of the alpha, beta, gamma and cosmic components of the radiation. The contribution of the alpha dose rate is normally eliminated by means of removing the outer 50 μm of the sample (i.e. the sample volume that has received external alpha radiation).

The beta dose rate decreases with depth and attenuation factors have to be taken into consideration. In addition, the beta component has to be calculated independently of the external gamma and cosmic dose rates. The external beta component is generated by the material (sediment or tissues) immediately in contact with the sample (2 mm range), whereas the external gamma dose rate is caused by all the sediment in a radius of 30 cm around the sample (figure 5.6). The external beta dose rate is calculated from the chemical analysis of U, Th and K performed in a laboratory, as shown previously. However, the gamma dose rate could be determined, either by analyzing the sediment surrounding the sample in the laboratory (using for example HpGe detectors) or directly '*in situ*' using a portable gamma spectrometer or thermoluminescence dosimeters. *In situ* gamma measurements show the advantage of including the present-day water contents. These measurements register the dose from all the surrounding environment (range 30 cm), and not only the dose from the sediment directly attached to the sample. In addition, data have to be adjusted using various correction factors in order to consider the water content (Grün 1994), as water absorbs beta and gamma radiation. Other parameters should also be considered for the corresponding corrections, such as the sample geometry, its thickness and density (see: Brennan *et al* 1991; Brennan 2003). Furthermore, it is required to assess the cosmic ray dose rate, which is around 300 μGy yr^{-1} at sea level and decreases with depth below ground, and it is also dependent on the altitude, latitude and the sediment nature.

The internal dose rate is mainly linked to the alpha and beta particles emitted from the radioelements within the sample. The self-absorption of internal gamma rays is normally negligible, considering the small thickness of most materials used for ESR dating, such as tooth enamel (when fossil teeth are dated) or quartz grains in sediments. For the calculation of the internal alpha component, it is required to consider that alpha particles are high energy and less efficient to produce ESR intensity than beta particles and gamma rays. Therefore, the alpha efficiency has to be determined, and is defined as the fraction of an equivalent beta and gamma dose that would yield the same ESR intensity (Aitken 1985; Grün 1989). Alpha efficiency values, for example in tooth enamel are in the range of 0.13 ± 0.02 (Grün and Katzenberger-Apel 1994).

In case of tooth enamel, only U-isotopes have to be taken into account for the calculation of the internal dose rate because tooth samples are usually free of K and Th. However, fossil teeth, corals or mollusc shells behave as open systems for uranium, which means that U could be incorporated or leached during the time the fossil was buried. Then, the uranium uptake history plays an important role in the calculation of the average dose rate. In the past, uranium uptake was normally assumed and three models were used for the calculation: the early U-uptake (EU), the linear U-uptake (LU) and the recent U-uptake (RU). In the EU model, it is assumed that all the uranium measured today was accumulated within the teeth in a short time after burial (Bischoff and Rosenbauer 1981). However, in the LU model, the incorporation of uranium is modeled to be constant over time (linear incorporation). Finally, in the recent uptake (RU) model, all measured U-concentration is described to be incorporated in recent times (Blackwell *et al* 1992). The differences among the EU, the LU and RU results are small if the uranium concentrations are relatively small. However, higher uranium concentrations cause important differences in the results, depending on the assumed model, EU, LU or RU. These parametric models allow easy calculations. However, they are based on unrealistic assumptions (i.e. the closed system assumption for the EU uptake) that simplify the complexity of the U-uptake in the different tooth tissues. These models lead to large uncertainties, especially when the uranium concentration within the dental tissues is high. To overcome the problem, the real U-uptake needs to be assessed and not assumed. Grün *et al* (1989) suggested to combine ESR and U-series analysis and introduced a parameter p to model the kinetics of the U-uptake in the dental tissues (US model). In comparison with the previous models, the combined US-ESR method has the advantage that the U-uptake is mathematically calculated (no assumed) considering the ESR and present-day U-series data measured in each tissue.

On the other hand, U-series disequilibrium in sediment is also possible to occur. Therefore, any U-series disequilibrium should be detected because of the significant impact on the dose rate calculation (Olley *et al* 1996).

It is necessary to highlight that the total dose rate evaluation is based on the analysis of present-day data extracted from the sample and its surrounding environment, and such data are considered to be representative of the past history. The data are assumed either having been constant over time or having followed a trend that could be known and accurately determined. Otherwise, any substantial modification of the sample or its surroundings, such as diagenetic processes: dissolution/recrystallization, recent U-mobilization, water circulation, etc, can introduce a bias into the dose rate assessment. A careful sampling is therefore required to minimize any potential source of error.

5.2.2.3 Age assessment

Once the D_E and the total dose rate have been calculated, the age assessment could be obtained from dividing the equivalent dose value by the dose rate ($T = D_E/D$). ESR ages are normally expressed at 1-sigma and final errors could reach values of around 10%–15%. This is mainly explained because of the high quantity of parameters (sometimes more than 20) that have to be considered and adjusted for the age calculation, such as: D_E, water content, radioelement content, alpha efficiency, gamma dose, density, cosmic dose, among others (see for instance: Grün 1989, 1992).

On the other hand, some samples require modeling (e.g. uranium in dental tissues), and the age assessment became more complex, and some iterative computations are required. At present, some non-commercial programs for ESR age calculations have been developed. For instance, DATA program (Grün 2009) and its Matlab equivalent US-ESR (Shao *et al* 2014) allow the age calculation using the combined U-series/ESR for teeth, and even corals or molluscs. For these two last materials is only required to adjust the alpha efficiency value. Other available programs offer similar possibilities, such as the software developed by Brennan *et al* (1999).

5.3 Potential of ESR dating and comparison with other chronometric dating methods

ESR dating shows an extraordinary potential for Quaternary geochronology and archaeology. One of the most interesting characteristics of ESR dating is that this method can theoretically cover the study of the whole Quaternary (last 2.6 Ma in any geological context). However, it is the high versatility the main advantage of ESR dating as ESR has been applied to a huge variety of materials (table 5.1), such as phosphates, silicates and carbonates, in different archaeological contexts since the pioneering study conducted by Ikeya (1975).

Since 1980 the most common materials that have been used in ESR dating include: speleothems, fossil teeth and optically bleached quartz grains. However, the interest for dating speleothems have been considerably reduced because of the

Table 5.1. Examples of ESR dating applications in archaeological contexts.

Material	Dated sample	Dated event	Example of application in an archaeological context
Carbonates	Speleothems	Formation of the speleothem	Valladas *et al* (2008)
	Mollusc shell	Shell deposition	Molodkov (2001) Blackwell *et al* (2012)
Phosphates	Human fossil teeth	Burial of the tooth	Grün (2006) Duval *et al* (2018)
	Animal fossil teeth	Death of the animal	Falguères *et al* (2006)
	Terrestrial apatite veins	Formation of the apatite veins	Rink *et al* (2003)
Flint/ Silicates	Sedimentary Quartz grains	Last sunlight exposure of the sediment	Voinchet *et al* (2010) Duval *et al* (2017)
	Heated quartz grains	Last time the quartz was heated	Monnier *et al* (1994)
	Faults (quartz)	Last fault movement	Ikeya *et al* (1982)
	Burnt Flint	Last time the flint was heated	Porat and Schwarcz (1991)
	Pottery	Last time the ceramic was heated	Bartoll and Ikeya (1997)

See Grün (1989), Ikeya (1993) and Rink (1997) for an extensive review of materials dated over the last 40 years.

development of mass spectrometry equipments that allow to provide U–Th dating results in a considerably faster and accurate way in comparison with ESR. However, when carbonates in karstic environments show ages higher than 600 ka, ESR can also be the preferred dating method.

It is also important to highlight that ESR dating is one of the few methods that can be applied to palaeontological remains, even hominid remains, allowing a direct dating process instead of the study of the sedimentary matrix. ESR dating is a valid option, especially when C-14 is not possible to be used for different reasons, such as the lack of collagen in the hominin bones (e.g. Torres *et al* 2010) or when samples from a period older than 50 ka are dated (Grün *et al* 2010). In addition, ESR dating of optically bleached quartz grains is a better option than OSL dating for samples older than 200 ka, which is usually beyond the OSL time range of application. ESR and OSL are therefore two dating methods that can be regarded as complementary for dating fluvial terraces, allowing to cover time ranges from Lower Pleistocene to Halocene (see for example Cordier *et al* 2012). Moreover, in Europe most of the archaeological sites are not found in volcanic contexts, but in karstic or alluvial environments, where the use of an absolute method such as Ar–Ar is just impossible.

On the other hand, the precision of the ESR dating process should be taken into account. ESR usually provides final errors between 10%–15% (1-sigma), although they could be reduced to values of 5%, in some particular situations. ESR and OSL show error values associated with the age results of the same magnitude order. These values are considerably higher than the ones provided by radiometric techniques, such as C-14, Ar–Ar or U/Th series, with high precision and error values <1% at 2-sigma. The results show that it is crucial to increase the precision of the ESR dating process. However, it should be taken into consideration that even if the development of ESR technology and methodologies experience a considerable improvement, it will be improbable to achieve final errors <5%. This is because radiometric methods are only based on measuring the radioactive decay of radionuclides contrary to the paleodosimetric methods (OSL, ESR, etc), which combine information about the sample and the environment. These methods normally require to take into account more than 15 variables along the whole analytical process (e.g. sample and environment water content, sample thickness and density, radionuclides concentrations, alpha efficiency, density, gamma dose, depth, latitude, etc) and each of these variables has an associated uncertainty, which should be considered for the final age calculation.

The time range for the application of ESR dating depends on the type of sample (in particular, of its paramagnetic centers nature) and the sample environment. For modern samples, the lower dating range is mainly characterized by two parameters: (a) the material radiation sensitivity and (b) the technological limitations. The latter refers to the equipment sensitivity employed for the dose reconstruction (specially the ESR spectrometer). As an example, recent studies in ESR dosimetry of tooth enamel lead to define a detection limit of several tens of mGy (Fattibene and Callens 2010). This small value suggests that ESR could be used for dating tooth enamel up to almost the present time. In the case of speleothems, the lower dating range has been assessed around 2 ka (Bassiakos 2001). Regarding the upper dating range, it is mainly dependent on the number of traps available in the crystal lattice (i.e. ESR signal saturation) and the thermal stability of the trap electrons. As previously described in section 5.2.1, the thermal stability of the ESR signal must be at least one order of magnitude higher than the age of the sample (Grün 2007). In this context, Schwarcz (1985) assesses the lifetime of the ESR signal of tooth enamel to around 10^9 at 25 °C, suggesting that ESR could be applied over a few Ma. In addition, for old tooth samples (>800 ka), the biggest concern is usually linked to the U-uptake modeling for the dose rate reconstruction process. In fact, U-leaching could be frequently observed for such time range and then, the application of the combined US-ESR approach becomes extremely difficult (e.g. Duval *et al* 2012b). Only two situations allow the possibility of using ESR to achieve accurate ages up to 2–3 Ma: (1) no apparent U-leaching is detected in dental tissues and (2) the tooth samples show a low U-concentration in dental tissues (e.g. Schwarcz *et al* 1994), which decrease the contribution of the dental tissues in the total dose rate calculation compared with the sediment.

In quartz grains, the ESR signal associated with the aluminum paramagnetic center, which is the most commonly used for quartz grains dating, shows almost no saturation with doses up to 60 000 Gy (Lin *et al* 2006) and is stable during approximately 7.4×10^9 (Toyoda and Ikeya 1991). This information makes possible the ESR dating process of quartz grains for the whole Quaternary.

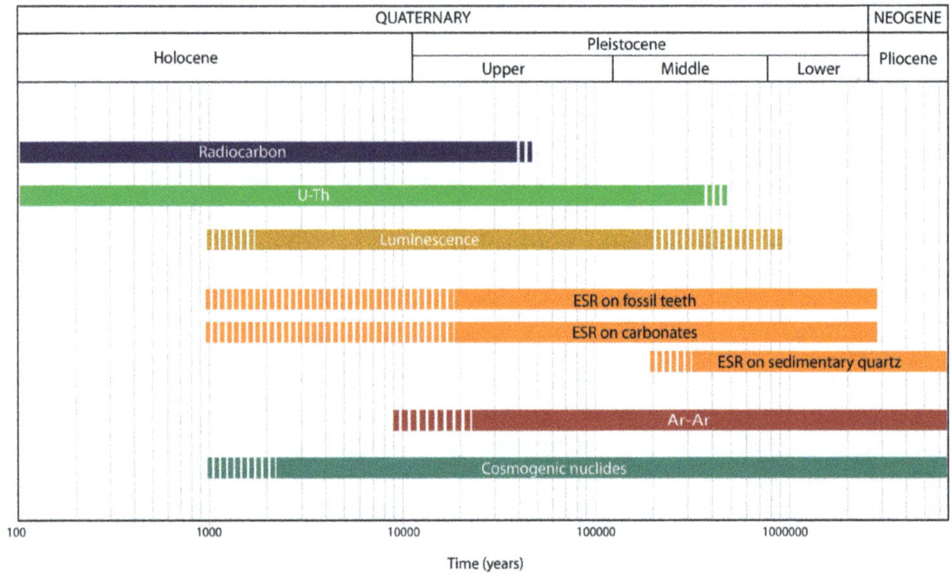

Figure 5.8. Some examples of chronometric dating methods with corresponding approximate time range of applicability. Time scale is logarithmic.

Although ESR dating competes with a great variety of other chronometric methods for the Middle and Late Pleistocene, it is not the same for the Early Pleistocene (figure 5.8). In fact, ESR dating has become the most widely paleodosimetric method used for the study of the first Hominin settlements in Europe (Falguères 2003; Bahain *et al* 2007; Duval *et al* 2012b). It is explained because of the different sedimentary contexts that limit the application of other absolute dating methods, such as OSL, Ar–Ar or cosmogenic nuclide dating. Therefore, further development of ESR dating is highly required in order to improve its accuracy and reliability. This will be crucial in the near future to complete the chronological frame of the first hominin settlements in Africa, Europe and Asia during the Early Pleistocene.

5.4 Examples of application of ESR dating in archaeology

ESR dating has been applied to a large variety of materials since the first work published by Ikeya (1975), and this versatility has become one of the most interesting advantages of ESR dating in the field of archaeology and quaternary science. This section will show a synthesis of some of the studies carried out in different sites in Europe, Middle East and North Africa (around the Mediterranean Sea) over the last 40 years. The section will be divided into several paragraphs according to the material type used in ESR dating. It will be mainly focused on fossil teeth and quartz grains as they have become the samples with better properties to achieve accurate results. In addition, other materials, such as corals, mollusc shells, speleothems and bones will be also briefly presented.

5.4.1 Tooth enamel

ESR dating of fossil teeth (mainly composed of hidroxyapatite mineral) is one of the most interesting materials because of its excellent dosimetric properties, becoming a highly relevant application of ESR dating. Fossil teeth allows direct dating of animals and hominid occupations from Early to Late Pleistocene time range. A summary of some of the most important archaeo-palaeontological sites of Europe and Middle East is shown in table 5.2.

Table 5.2. List of some archaeo-palaeontological sites of Europe and Middle East dated by means of electron spin resonance (ESR) using tooth enamel.

Sites	Application	Time range (ka)	Reference
Sima de los Huesos, Atapuerca, Spain	Hominins Homo heidelbergensis	400–500	Aguirre *et al* (1990) Bischoff *et al* (1997) Arsuaga *et al* (2014)
Piégu, France	Middle Palaeolithic Flake Industry	100–113	Bahain *et al* (1993)
Aragó, France	Hominids	200–355	Bahain *et al* (1993)
Fuente Nueva-3, Spain	Mammoth Mode 1 lithics industry	1000–1200	Duval *et al* (2012a)
Valparadís, Spain	Mode 1 lithic industry	660–860	Duval *et al* (2015)
Caverna delle Fate, Italy	Hominids	60–74	Falguères *et al* (1990)
Ambrona, Spain	Acheulian lithics	284–316	Falguères *et al* (2006)
Stuttgart, Germany	Mammoth		Grün and Brunnacker (1987)
Tabun Cave, Israel	Neanderthals, Modern Humans, Lower and Middle Palaeolithic lithics	80–260	Grün *et al* (1991)
Amud Cave, Israel	Neanderthals, Levallois Mousterian lithics	40–50	Grün and Stringer (1991)
La Chapelle-aux-Saints, France	Neanderthals and Quina Mousterian lithics	47–56	Grün and Stringer (1991)
Banyolas, Spain	Neanderthals	50–60	Grün *et al* (2006)
Misliya Cave, Israel	Hominins	170–180	Hershkovitz *et al* (2018)
Abauntz, Spain	Lithic industry	25–30	Mazo *et al* (2012)
Skhul, Tabun, Qafzeh, Israel	Hominids and Lithics	40–160	McDermott *et al* (1993)
Krapina, Croatia	Neanderthals and Mousterian lithics	113–130	Rink *et al* (1995)

Kulna, Czech Republic	Neanderthals and Micoquian lithics	46–50	Rink *et al* (1996)
Grotte Breuil, Italy	Neanderthal, Late Middel Paleolithic lithics		Schwarcz *et al* (1991)
Kebara, Israel	Neanderthal, Mousterian lithics	60–64	Schwarcz *et al* (1989)
Hoxne, England	Acheulian lithics		Schwarcz and Grün (1993)
Qafzeh, Israel	Modern Humans, Middle Palaeolithic lithics	96–115	Schwarcz *et al* (1988)

The published results suggest that ESR dating allows to cover an extense time range when using fossil tooth enamel. There are some studies that cover ages from around only 25–30 ka, such as the one conducted by Mazo *et al* (2012) and others that focus on very old remains, even older than 1 million years (Duval *et al* 2012a). In addition, it is worth highlighting that ESR could be applied to different environments, including not only open-air sites but also karstic caves.

The first ESR dating examples with mammalian teeth in archaeological sites were successful through intercomparison with thermoluminesce dating of burnt flint at Qafzeh and Kebara sites in Israel (Valladas *et al* 1987). Then, many archaeologists and paleanthropologists started to accept that anatomically modern humans and Neanderthals had lived at the same time. At that moment, ESR dating applications were widespread to study Human Evolution due to the close spatial relationships between mammal teeth and human fossils (Grün and Stringer 1991). Nowadays, Human Evolution studies have become one of the most important fields of research. However, finding a tooth, together with other cultural materials was quite common. Therefore, the method applications were extended into the area of dating cultural materials, such as stone tool assemblages (Rink *et al* 1996), bone flutes (Lau *et al* 1997), among others.

The particular strength of ESR dating is the virtually non-destructive, direct analysis of tooth enamel fragments from hominid fossils (direct dating, Grün *et al* 2006). Torres *et al* (2010) is a relevant study that combines various chronometric dating methods and includes the dating of a fragment of human incisor from El Sidrón Cave in Spain. The tooth material was dated by a semi-destructive analytical procedure based on the combination of ESR analysis with LA-ICP-MS U-series analysis. The ESR age estimation of 38.5 ± 4.5 ka was found to be in good agreement with the results derived from the other dating methods, such as radiocarbon, U-series, OSL, and amino acid racemization, which ranged from 30 to 50 ka for the archaeological unit (see for further details in Torres *et al* 2010). Geological occurrences of hydroxyapatite can also be dated, as illustrated by the work performed on authigenic terrestrial apatite veins at Tabun Cave in Israel, which provided useful information about the diagenetic processes that had occurred in the deposits (Rink *et al* 2003).

5.4.2 Quartz grains and other silicates

Silicates, and especially quartz minerals, are also widely used for ESR dating. The most common application with quartz corresponds to ESR dating of optically bleached quartz grains extracted from sediments (table 5.3), same application as OSL dating.

Quartz is a ubiquitous mineral that is found in almost any geological context, showing a great potential for dating. The first ESR dating attempt of optically bleached quartz grains extracted from Quaternary sediment was published more than 30 years ago (Yokoyama *et al* 1985). Since then, a few research groups based in France, Germany, Spain, Canada, China or Japan have been working on this application over the last decades. ESR analysis of quartz has been used for dating river terraces (for instance, Voinchet *et al* 2004) and even cave sediments (Yokoyama *et al* 1985; Parés *et al* 2018: table 5.3).

Table 5.3. List of some archaeo-palaeontological sites in Europe and Middle East dated by electron spin resonance (ESR) using quartz grains.

Sites	Archaeological/ Palaeontological site material	Time range (ka)	Reference
Cuesta de la Bajada, Spain	Lithic Industry	264–451	Santonja *et al* (2014)
Vallparadís, Spain	Mode 1 lithic industry	778–1046	Duval *et al* (2015)
Barranco León, Spain	Mode 1 lithic industry and Hominids	1020–1730	Duval *et al* (2012a); Toro-Moyano *et al* (2013)
Ain Boucherit, Algeria	Oldowan stone artifacts and Hominins	1900–2400	Sahnouni *et al* (2018)
Porto Maior, Spain	Acheulean tools	206–279	Mendez-Quintas *et al* (2018).
Ubeidiya, Israel	Early Acheulian artifacts and Fauna	1200–1300	(Rink *et al* 2007)
Arago cave and Terra Amata, France	Aucheulian	380–430	Yokoyama *et al* (1985)
Ca' Belvedere di Monti Poggiolo Italy	Palaeolithic artefacts	900–1100	Gagnepain *et al* (1998)
Pont-de-Lavaud, France	Lithic industry	1000–1100	Voinchet *et al* (2004); Despriée *et al* (2006)
Somme basin, France	Acheulin and Mousterian sites with fauna	270–1870	Laurent *et al* (1994)
Tourville-la-Rivière, France	Mammal Fauna and Lithic tools	236–243	Bahain *et al* (2019)
Gran Dolina, Atapuerca, Spain	Hominin remains and lithic artifacts	840–2300	Parés *et al* (2018)

The main advantages of ESR over luminescence methods are that the Al and Ti centers in quartz have significantly higher saturation levels than most luminescence methods. This means that old samples are able to be datable. Therefore, ESR is one of the few dating methods that can be applied to Early Pleistocene sites. For example, ESR was applied to both quartz grains and fossil teeth at the archaeological site of Vallparadís in Spain (Duval *et al* 2015, and references therein). Both applications yielded consistent results around 0.8 Ma for the main archaeological level. In addition, the mean ESR age of 858 ± 87 ka was in excellent agreement with the results derived from other techniques: palaeomagnetism and biochronology. Considering all those results, it is suggested a Matuyama age for the deposits, and more precisely between the Brunheschron and the Jaramillo subchron (i.e. 0.99–0.78 Ma). In recent years, different intercomparison studies have been carried out using several dating methods, such as OSL, ESR, U-series, with the aim of comparing the results obtained with the different dating methods. This is of special interest because it will allow to provide reliable and confirmed chronologies (Duval *et al* 2017; Kreutzer *et al* 2018; Parés *et al* 2018).

Other applications have been also published. Quartz is a useful mineral for ESR dating. Fault movements, tephras, sediments baked by lava flow, burned flints have been also dated using quartz. For example, Monnier *et al* (1994) published ESR chronologies based on heated quartz grains extracted from fireplace sediments and a burnt quarzite pebble to study the lower palaeolithic site of Menezdregan in France. Porat and Schwarcz (1991) showed the potential of burnt flint for dating and some samples from Middle Eastern archaeological sites were studied. Moreover, in the same way as thermoluminescence (TL) method, ESR could be used to date pottery firing. Bartoll and Ikeya (1997) analyzed fragments from a Neolithic vessel excavated in Rohrback, Germany, which showed ESR ages of ~5 ka. These results were in agreement with the TL ages obtained on the same fragments.

Other authors, such as Grün, have applied ESR for the reconstruction of the thermal history of the Eddzhurtinskiy Granite, Caucasus (Grün *et al* 1999). The study suggested that ESR may be applicable for detailed landscape reconstructions of rapidly uplifting mountains terrains.

5.4.3 Speleothems

Speleothems are secondary carbonates precipitated in caves (stalagmites, stalactites, and flowstones). These deposits are often connected with archaeological sites and also contain paleoclimatic information. Therefore, speleothems have been studied for dating aims by different methods. The first dating attempt using ESR was carried out on a stalactite from Akiyoshi Cave in Japan (Ikeya 1975) and it was the oldest application of ESR method for dating. According to Grün (1989) and Blackwell (1995) more than 40 examples regarding ESR speleothems dating were reported. In addition, it is also known some recent applications, such as the one published by Valladas *et al* (2008). The authors obtained ESR age estimates using a stalagmitic formation from the Neanderthal Payre Cave in France and the late Middle Pleistocene ages results (<300 ka) were consistent with the corresponding U-series ages studied on the same unit. Although it is possible to use ESR for dating speleothems, U-series is much more precise and requires less sample and effort, particularly with the recent advances of ICP multicollector mass spectrometry and *in situ* laser ablation analysis (Potter *et al* 2005; Eggins *et al* 2005). Therefore, U-series dating of secondary carbonates is now much faster and cheaper than ESR, and nowadays ESR dating is especially used for old samples, when carbonates in karstic environments show ages higher than 600 ka.

5.4.4 Corals and shells

ESR dating has been useful in sorting out the chronology of the raised coral reef tracts in Barbados (Schellmann and Radtke 2004). The ESR age estimates agree well with the radiocarbon results in the range of 0–3500 years, and both methods have similar-sized errors. Compared to U-series results in the range of 70 000–130 000 years, ESR seems to provide results that are somewhat younger than the expected ages and the individual errors are significantly larger than those of U-series. U-series seems to have a trend of slightly overestimating the expected age. One has to keep in mind that ESR seems significantly less affected than U-series by diagenetic processes (Eggins *et al* 2005).

Regarding shells, Molodkov (2001) dated several terrestrial shell samples in Treugolnaya Cave in Russia, from deposits associated with Early Acheulian lithic artifacts. Molodkov achieved ESR ages older than 350 ka. In a similar way, Blackwell *et al* (2012) dated mollusc shells from a Late Pleistocene archaeological event in the Western Desert in Egypt. To sum up, these applications suggest that ESR could potentially cover a wide time ranges, from recent samples, even with only 5 ka to materials as old as >2 Ma (Molodkov 2001; Blackwell *et al* 2012). However, their reliability, as in the case of fossil teeth, depends on the accuracy of the U-uptake modeling.

5.4.5 Bones

Although it would be highly desirable to date bones with ESR, all attempts to obtain reliable ESR age estimates have not been successful. This is due to the following observations: bones usually absorb more uranium than teeth, and the mineral phase in bones (hydroxyapatite) is only 40%–60% (enamel, ~96%). During fossilization, the mineralogic compounds of the bones change, leading to the formation of new minerals, disintegration of some mineral phase, conversion of an amorphous phase into hydroxyapatite, as well as growth of the crystal size. Formation of the new mineral phase (with new defects = traps) with time will generally lead to age underestimations regardless of the U-uptake model applied (Grün 2007).

5.5 Future scope and key issues

ESR dating, as any other dating method, show strengths and limitations in the field of archaeology. Although ESR dating is not able to reach the same precision than radiometric methods (C-14, Ar–Ar, U–Th), ESR is probably the only dating method that can be potentially used in almost any geological context. Therefore, ESR has become the only possible dating method in some contexts and time ranges, thanks to its high versatility. For example, ESR dating has allowed the study of the first hominid settlements in Europe, such as Atapuerca, Fuente Nueva-3 and Barranco León sites.

The main advantage of ESR dating is to allow the direct dating of hominid remains beyond the applicability of radiocarbon (i.e. C-14 limited to 50 ka). Unfortunately, bones are not suitable for age assessments by ESR dating. This is because complex diagenetic processes occur during burial, leading to the destruction of some bone structures. However, fossil enamel appears to be suitable for direct dating with ESR. In the last 40 years, many studies have been carried out on tooth enamel. The successful application on archaeological and palaeoanthropological sites has been reported for fossils ranging from a few thousand years (e.g. Grün *et al* 1996) to about a million years (e.g. Han *et al* 2012) and exceptionally to a few millions of years (Curnoe *et al* 2001). However, a strong limitation is associated with the destructive nature of ESR dating of fossil remains. Fossil enamel was usually crushed into powder; nevertheless, recent developments in ESR dating allows the direct non-destructive dating of fossil enamel fragments (Joannes-Boyau and Grün 2009, 2011). The first step of sample preparation requires to remove a fragment from the tooth. The use of natural cracks in the enamel offers the best opportunity to separate the fragment, causing minimal damage to the remaining tooth. Small fractures frequently occur on fossil tooth enamel during burial caused by environmental processes. When the analysis has been completed, the fragment can be reattached to the tooth, with little visible damage. As the impact on the fragment itself is minimal, this protocol is considered to be virtually non-destructive. In addition, new developments in the analytical technique, such as the study of small enamel fragment using high resolution U–Th analysis in combination with LA-ICP-MS (Laser Ablation Inductively Coupled Plasma Mass Spectrometry) and ESR allows to reduce the mass required for the analysis considerably. Therefore, in the last years LA-ICP-MS has become a key tool in the analytical procedure of fossil teeth dating. It is possible to scan the teeth and map the spatial distribution of uranium series elements in different dental tissues (see for example, Duval *et al* 2011). This is quite relevant because it helps to identify the best areas for ESR dating. In addition, this procedure allows to remove the inadequate samples easily (for instance, samples with a high uranium concentration in the enamel or samples that have suffered from uranium leaching).

Other materials, carbonate crystal, such as corals, speleothems, calcites, travertines, and mollusc shells, can be dated by ESR; however, the complexity of the signal and doses calculation constrained severely the accuracy. U-series dating and radiocarbon dating are usually the preferred methods, especially for corals and speleothems, unless time range is above U–Th series application (around 600 ka).

Optically bleached quartz grains show an extraordinary chronological interest because quartz is the most common mineral in the earth surface. Then, with the increasing number of archaeological sites in fluvial context (Mishra *et al* 2007), ESR could play a key role for the establishment of accurate chronologies, especially for periods beyond the standard OSL application (>200 ka). Then ESR has shown an enormous potential for dating Middle and Early Pleistocene sites, especially in alluvial contexts.

On the other hand, the progress of microwave technology and the development of the ESR dating protocols during the last 20 years have aimed to minimize the amount of sample required for the dating process. Therefore, some new ESR dating methods, such as the use of the single aliquot procedure, performing measurements of single crystals, have been developed. Other examples are based on the use of ESR spectrometers with higher frequency bands, specially Q-band spectrometers ($\upsilon \sim 34$ GHz), with the objective to increase the spectral resolution and sensitivity of the measurements dramatically (Beerten *et al* 2003; Guilarte *et al* 2016). As a consequence, Q-band spectrometers reduce the sample mass to values of only a few milligrams (<5 mg, De *et al* 2013; Romanyukha *et al* 2007).

In addition, new methodological and technical developments in ESR dating with the aim of improving accuracy and reliability of the method will be highly relevant in the near future. This new progress will be crucial to complete and improve the chronological frame of the first hominin settlements in Africa, Europe and Asia during the Early Pleistocene.

References

Aguirre E *et al* 1990 The Atapuerca sites and the ibeas hominids *Hum. Evol.* **5** 55–73

Aitken M J 1985 *Thermoluminescence Dating* (London: Academic)

Ankjærgaard C and Murray A S 2007 Total beta and gamma dose rates in trapped charge dating based on beta counting *Radiat. Meas.* **42** 352–9

Arnold L J, Duval M, Falguères C, Bahain J-J and Demuro M 2012 Portable gamma spectrometry with cerium-doped lanthanum bromide scintillators: suitability assessments for luminescence and electron spin resonance dating applications *Radiat. Meas.* **47** 6–18

Arsuaga J L *et al* 2014 Neandertal roots: cranial and chronological evidence from Sima de los Huesos *Science* **344** 1358–63

Bahain J J, Falguères C and Yokoyama Y 1993 Attempt at ESR dating of tooth enamel of French Middle Pleistocene sites *Appl. Radiat. Isot.* **43** 267–72

Bahain J-J, Falguères C, Voinchet P, Duval M, Dolo J-M, Despriée J, Garcia T and Tissoux H 2007 Electron Spin resonance (ESR) dating of some European Late Lower Pleistocene sites *Quaternaire* **18** 175–86

Bahain J J *et al* 2019 ESR and ESR/U-series chronology of the Middle Pleistocene site of Tourville-la-Rivière (Normany, France)—A multi laboratory approach *Quat. Int.* **556** 66–78

Bartoll J and Ikeya M 1997 ESR dating of pottery: a trial *Appl. Radiat. Isot.* **48** 981–4

Bassiakos Y 2001 Assessment of the lower ESR dating range in Greek *speleothems J. Radioanal. Nucl. Chem.* **247** 629–33

Beerten K, Pierreux D and Stesmans A 2003 Towards single grain ESR dating of sedimentary quartz: first results *Quat. Sci. Rev.* **22** 1329–34

Bischoff L and Rosenbauer R J 1981 Uranium-series dating of human skeletal remains from the Del Mar and Sunnyvale sites, California *Science* **213** 1003–5

Bischoff J L, Fitzpatrick J A, León L, Arsuaga J L, Falgueres C, Bahain J-J and Bullen T 1997 Geology and preliminary dating of the hominid-bearing sedimentary fill of the Sima de los

Huesos Chamber, Cueva Mayor of the Sierra de Atapuerca, Burgos, Spain *J. Hum. Evol.* **33** 129–54

Blackwell B, Porat N, Schwarcz H P and Debenath A 1992 ESR dating of tooth enamel: comparison with $^{230}Th/^{234}U$ speleothem dates at La Chaise-de-Vouthon (Charente), France *Quat. Sci. Rev.* **11** 231–44

Blackwell B A 1995 Electron spin resonance dating *Dating Methods for Quaternary Deposits Geotext 2* ed N W Rutter and N R Catto (Canada: Geological Association of Canada) pp 209–68

Blackwell B A B, Skinner A R, Mashriqi F, Deely A E, Long R A, Gong J J J, Kleindienst M R and Smith J R 2012 Challenges in constraining Pluvial events and Hominin activity: examples of ESR dating Mollucs from the Western Desert, Egypt *Quat. Geochronol.* **10** 430–5

Brennan B J, Lyons R G and Phillips S W 1991 Attenuation of alpha particle track dose for spherical grains *Nucl. Tracks Radiat. Meas.* **18** 249–53

Brennan B J, Rink W J, Rule E M, Schwarcz H P and Prestwich W V 1999 The ROSY ESR dating program *Anc. TL* **17** 45–53

Brennan B J 2003 Beta doses to spherical grains *Radiat. Meas.* **37** 299–303

Cordier S, Harmand D, Lauer T, Voinchet P, Bahain J-J and Frechen M 2012 Geochronological reconstruction of the Pleistocene evolution of the Sarre valley (France and Germany) using OSL and ESR dating techniques *Geomorphology* **165–166** 91–106

Curnoe D, Grün R, Taylor L and Thackeray F 2001 Direct ESR dating of a Pliocene hominin from Swartkrans *J. Hum. Evol.* **40** 379–91

De T, Romanyukha A, Trompier F, Pass B and Prabhakar M 2013 Feasibility of Q-band EPR dosimetry in biopsy samples of dental enamel, dentine and bone *Appl. Magn. Reson.* **44** 375–87

Despriée J, Gageonnet R, Voinchet P, Bahain J J, Falguères C, Varache F, Courcimault G and Dolo J M 2006 Une occupation humaine au Pléistocene inférieur sur la bordure nord du Massif central *C. R. Palevol.* **5** 821–8

Duval M, Grün R, Falguères C, Bahain J-J and Dolo J-M 2009 ESR dating of lower Pleistocene fossil teeth: limits of the single saturating exponential (SSE) function for the equivalent dose determination *Radiat. Meas.* **44** 477–82

Duval M, Aubert M, Hellstrom J and Grün R 2011 High resolution LA-ICP-MS mapping of U and Th isotopes in an early Pleistocene equid tooth from Fuente Nueva-3 (Orce, Andalusia, Spain) *Quat. Geochronol.* **6** 458–67

Duval M and Guilarte V 2012 Assessing the influence of the cavity temperature on the ESR signal of the Aluminumcenter in quartz grains extracted from sediment *Anc. TL* **30** 11–6

Duval M *et al* 2012a On the limits of using combined U-series/ESR method to date fossil teeth from two Early Pleistocene archaeological sites of the Orce area (Guadix-Baza basin, Spain) *Quat. Res.* **77** 482–91

Duval M, Falguères C and Bahain J-J 2012b Age of the oldest hominin settlements in Spain: Contribution of the combined U-series/ESR dating method applied to fossil teeth *Quat. Geochronol.* **10** 412–7

Duval M and Arnold L J 2013 Field gamma dose-rate assessment in natural sedimentary contexts using $LaBr_3$(Ce) and NaI (Tl) probes: a comparison between the ‘threshold’ and ‘windows’ techniques *Appl. Radiat. Isot.* **74** 36–45

Duval M, Bahain J-J, Falguères C, Garcia J, Guilarte V, Grün R, Martínez K, Moreno D, Shao Q and Voinchet P 2015 Revisiting the ESR chronology ofthe Early Pleistocene hominin occupation at Vallparadís(Barcelona, Spain) *Quat. Int.* **389** 213–23

Duval M, Arnold L, Guilarte V, Demuro M, Santonja M and Pérez-González A 2017 Electron Spin Resonance dating of optically bleached quartz grains from the Middle Palaeolithic site of Cuesta de la Bajada (Spain) using the multiple centres approach *Quat. Geochronol.* **37** 82–96

Duval M, Grün R, Parés J M, Martín-Francés L, Campaña I, Rosell J, Shao Q, Arsuaga J L, Carbonell E and Bermúdez de Castro J M 2018 The first direct ESR analysis of a hominin tooth from Atapuerca Gran Dolina TD-6 (Spain) supports the antiquity of Homo antecessor *Quat. Geochronol.* **47** 120–37

Eaton G R, Eaton S S, Barr D P and Weber R T 2010 *Quantitative EPR* (New York: Springer)

Eggins S *et al* 2005 In siru U-series daring by laser-ablarionmulri-collecror ICPMS: New prospects for Quaternary geochronology *Quat. Sci. Rev.* **24** 2523–38

Falguères C, Yokoyama Y and Eibron R 1990 Electron Spin Resonance (ESR) dating of hominid-bearing deposits in the Cavernadelle Fate, Ligure, Italy *Quat. Res.* **37** 121–8

Falguères C 2003 ESR dating and the human evolution: contribution to the chronology of the earliest humans in Europe *Quat. Sci. Rev.* **22** 1345–51

Falguères C, Bahain J-J, Pérez-González A, Mercier N, Santonja M and Dolo J-M 2006 The lower Acheulian site of Ambrona, Soria (Spain): ages derived from a combined ESR/U-series model *J. Archaeol. Sci.* **33** 149–57

Fattibene P and Callens F 2010 EPR dosimetry with tooth enamel: a review *Appl. Radiat. Isot.* **68** 2033–116

Gagnepain J, Laurent M, Bahain J J, Falguères C, Hedley I, Peretto C, Wagner J J and Yokoyama Y 1998 Synthèse des données paléomagnétiques et radiochronologiques du site de Ca' Belvedere di Monte Poggiolo (Romagna, Italie) et de son environnement géologique *XIII Congresso delle Scienze Preistoriche e Protostoriche Workshop 13 Forlì 1999* vol 6 (Abaco: Abaco Editore) pp 877–88

Goldstein S J and Stirling C H 2003 *Techniques for Measuring Uranium-series Nuclides: 1992–2002* vol 52 ed B Bourdon, G M Henderson, C C Lundstrom and S P Turner (Washington: Mineralogical Society of America) pp 23–58

Grün R 1989 Electron spin resonance (ESR) dating *Quat. Int.* **1** 65–109

Grün R 1992 Suggestions for minimum requirements for reporting ESR age estimates *Anc. TL* **10** 37–41

Grün R 1994 A cautionary note: use of 'water content' and 'depth for cosmic ray dose rate' in AGE and DATA programs *Anc. TL* **12** 50–1

Grün R 2006 Direct dating of human fossils *Yearb. Phys. Anthropol.* **49** 2–48

Grün R 2007 *Encyclopeda of Quaternary Science* ed S A Elias (Amsterdam: Elsevier) pp 1505–16

Grün R 2009 The DATA program for the calculation of ESR age estimates on tooth enamel *Quat. Geochronol.* **4** 231–2

Grün R and Brunnacker K 1987 ESR DatierungeinesMammuth-ZahnesausdemTravertinbruch 'Lauster', Stuttgart-Bad Canstatt *Jhrb. Geol. Landes-Baden-Würtemburg* **28** 43–7

Grün R, Stringer C B and Schwarcz H P 1991 ESR dating of teeth from Garrod's Tabun Cave collection *J. Hum. Evol.* **20** 231–48

Grün R and Stringer C B 1991 Electron spin resonance dating and the evolution of modern humans *Archaeometry* **33** 153–99

Grün R and Katzenberger-Apel O 1994 An irradiator for ESR dating *Anc. TL* **12** 35–8

Grün R, Brink J S, Spooner N A, Taylor L, Stringer C B, Franciscus R G and Murray A S 1996 Direct dating of Florisbad hominid *Nature* **382** 500–1

Grün R, Tani A, Gurbanov A, Koshchug D, Williams I and Braun J 1999 A new merhod for the estimarion of cooling and denudarion rates using paramagnetic cenrres in quartz: A case study on the Eldzhurtinskiy Granite, Caucasus *J. Geophys. Res.* **104** 17531–49

Grün R, Maroto J, Eggins S, Stringer C, Robertson S, Taylor L, Mortimer G and McCulloch M 2006 ESR and U-series analyses of enamel and dentine fragments of the Banyoles mandible *J. Hum. Evol.* **50** 347–58

Grün R, Aubert M, Hellstrom J and Duval M 2010 The challenge of direct dating old human fossils *Quat. Int.* **223–224** 87–93

Guérin G, Mercier N and Adamiec G 2011 Dose-rate conversion factors: update *Anc. TL* **29** 5–8

Guilarte V and Duval M 2020 ESR dating of optically bleached quartz grains: intra-laboratory comparison of different experimental setups and their impact on dose evaluation *Geochronometria* **48** 179–90

Guilarte V, Trompier F and Duval M 2016 Evaluating the Potential of Q-band ESR Spectroscopy for Dose Reconstruction of Fossil Tooth Enamel *PLoS One* **11** e0150346

Han F *et al* 2012 Preliminary results of combined ESR/U-series dating of fossil teeth from Longgupo cave, China *Quat. Geochronol.* **10** 436–42

Hershkovitz I *et al* 2018 The Earliest Modern Human outside *Africa Science* **359** 456–45

Ikeya M 1975 Dating a stalactite by electron paramagnetic resonance *Nature* **255** 48–50

Ikeya M, Miki T and Tanaka K 1982 Dating a fault by Electron Spin Resonance on intrafault materials *Science* **215** 1392–3

Ikeya M 1993 *New Applications of Electron Spin Resonance - Dating, Dosimetry and Microscopy* (Singapore: World Scientific)

Joannes-Boyau R and Grün R 2009 Thermal behavior of oriented and non-oriented CO_2 radicals in tooth enamel *Radiat. Meas.* **44** 505–11

Joannes-Boyau R and Grün R 2011 A comprehensive model for CO_2- radicals in fossil tooth enamel: Implications for ESR dating *Quat. Geochronol.* **6** 82–97

Kreutzer S, Duval M, Bartz M, Bertran P, Bosq M, Eynaud F, Verdin F and Mercier N 2018 Deciphering Long-Term Coastal Dynamics using IR-RF and ESRDating: A Case Study from Médoc, South-West France *Quat. Geochronol.* **48** 108–20

Lau B, Blackwell B, Schwarcz H P, Turk I and Blicktein J I 1997 Dating a flautist? Using ESR (Electron Spin Resonance) in the Mousterian cave deposits at Divje Babe I, Slovenia *Geoarchaeology* **12** 507–36

Laurent M, Falguères C, Bahain J J and Yokoyama Y 1994 Geochronologie du système de terrasses fluviatiles quaternaires du bassin de la Somme par datation RPE sur quartz, déséquilibres des familles de l'uranium et magnétostratigraphie *C. R. Acad. Sci., Paris* **318** 521–6

Lin M, Yin G, Ding Y, Cui Y, Chen K, Wu C and Xu L 2006 Reliability study on ESR dating of the aluminumcenter in quartz *Radiat. Meas.* **41** 1045–9

Mazo C, Utrilla P, Blasco F, Mandado J, Torres T, Ortiz E and Rink J W 2012 El nivel musteriense de la cueva de Abauntz (Arraitz, Navarra) y su aportación al debate 'vasconiense' *Mainake* **33** 187–214

McDermott F, Grün R, Stringer C B and Hawkesworth C J 1993 Mass-spectrometric U-series dates for Israeli Neanderthal/early modern hominid sites *Nature* **363** 252–5

Mendez-Quintas E, Santonja M, Pérez-González A, Duval M, Demuro M and Arnold L 2018 First evidence of an extensive Acheulean large cutting tool accumulation in Europe from Porto Maior (Galicia, Spain) *Sci. Rep.* **8** 3082

Miallier D, Guérin G, Mercier N, Pilleyre T and Sanzelle S 2009 The Clermont radiometric reference rocks: a convenient tool for dosimetric purposes *Anc. TL* **27** 37–43

Mishra S *et al* 2007 Fluvial deposits as an archive of early human activity *Quat. Sci. Rev.* **26** 2996–16

Molodkov A 2001 ESR dating evidence for early man at a Lower Palaeolithic cave-site in the Northern Caucasus as derived from terrestrial mollusc shells *Quat. Sci. Rev.* **20** 1051–5

Monnier J L *et al* 1994 A new regional group of the Lower Palaeolithic in Brittany (France), recently dated by Electron Spin Resonance *C. R. Acad. Sci.* **319** 155–60

Olley J M, Murray A and Roberts R G 1996 The effects of disequilibria in the uranium and thorium decay chains on burial dose rates in fluvial sediments *Quat. Sci. Rev.* **15** 751–60

Parés J M *et al* 2018 Chronology of the cave interior sediments at Gran Dolina archaeological site, Atapuerca (Spain) *Quat. Sci. Rev.* **186** 1–16

Porat N and Schwarcz H P 1991 Use of signal subtraction methods in ESR dating of burned flint *Nucl. Tracks Radiat. Meas.* **18** 203–12

Potter E K, Stirling C H, Wiechert U H, Halliday A N and Spötl C 2005 Uranium-series dating of corals *in situ* using laser-ablation MC-ICPMS *Int. J. Mass spectrom.* **240** 27–35

Regulla D F, Wieser A and Göksu H Y 1985 Effects of sample preparation on the ESR spectra of calcite, bone and volcanic material *Nucl. Tracks* **10** 825–30

Rink W J, Schwarcz H P, Smith F H and Radovcic J 1995 ESR dating of Krapina hominids *Nature* **378** 24

Rink W J, Schwarcz H P, Valoch K, Seitl L and Stringer C B 1996 Dating of Micoquian Industry and Neanderthal remains at Kulna, Czech Republic *J. Archaeol. Sci.* **23** 889–901

Rink W J 1997 Electron spin resonance (ESR) dating and ESR applications in quaternary science and archaeometry *Radiat. Meas.* **27** 975–1025

Rink W J, Bartoll J, Goldberg P and Ronen A 2003 ESR dating of archaeologically relevant authigenic terrestrial apatite veins from Tabun Cave, Israel *J. Archaeol. Sci.* **30** 1127–38

Rink W J, Bartoll J, Schwarcz H P, Shane P and Bar-Yosef O 2007 Testing the reliability of ESR dating of optically exposed buried quartz sediments *Radiat. Meas.* **42** 1618–26

Romanyukha A, Mitchell C A, Schauer D A, Romanyukha L and Swartz H M 2007 Q-band EPR biodosimetry in tooth enamel microsamples: feasiblility test and comparison with X-band *Health Phys.* **93** 631–5

Sahnouni M *et al* 2018 1.9-Million- and 2.4-Million-year-old artifacts and stone tool-cutmarked bones from Ain Boucherit, Algeria *Science* **362** 1297–301

Santonja M *et al* 2014 The Middle Paleolithic site of Cuesta de la Bajada (Teruel, Spain): a perspective on the Acheulean and Middle Paleolithic technocomplexes in Europe *J. Archaeol. Sci.* **49** 556–71

Schellmann G and Radtke U 2004 The marine Quarernary of Barbados *Kain. Geogr. Arb.* **81** 1–137

Schwarcz H P 1985 ESR study of tooth enamel *Nucl. Tracks* **10** 865–7

Schwarcz H P, Grün R, Vandermeersch B, Bar-Yosef O, Valladas H and Tchernov E 1988 ESR dates for the hominid burial site of Qafzeh in Israel *J. Hum. Evol.* **17** 733–7

Schwarcz H P, Buhay W M, Grün R, Valladas H, Tchernov E, Bar-Yosef O and Vandermeersch B 1989 ESR dating of the Neanderthal site of Kebara Cave, Israel *J. Archaeol. Sci.* **16** 653–9

Schwarcz H P, Buhay W, Grün R, Steiner M, Kuhn S and Miller G H 1991 Absolute dating of sites in coastal Lazio *Quat. Nova* **1** 51–67

Schwarcz H P and Grün R 1993 Electron Spin Resonance (ESR) dating of the Lower Industry at Hoxne *The Lower Paleolithic Site at Hoxne* ed B G Gladfelter and J Wymer (London: University of Chicago Press) pp 210–1

Schwarcz H P, Grün R and Tobias P V 1994 ESR dating studies of the australopithecine site of Sterkfontein, South Africa *J. Hum. Evol.* **26** 175–81

Shao Q, Bahain J-J, Dolo J-M and Falguères C 2014 Monte Carlo approach to calculate US-ESR age and age uncertainty for tooth enamel *Quat. Geochronol.* **22** 99–106

Toro-Moyano I *et al* 2013 The oldest human fossil in Europe, from Orce (Spain) *J. Hum. Evol.* **65** 1–9

Toyoda S and Ikeya M 1991 Thermal stabilities of paramagnetic defects and impurity centers in quartz: basis for ESR dating of thermal history *Geochem. J.* **25** 437–45

Torres T *et al* 2010 Dating of the hominid (Homo neanderthalensis) remains accumulation from el Sidrón cave (Piloña, Asturias, North Spain): an example of a multi-methodological approach to the dating of upper Pleistocene sites *Archaeometry* **52** 680–705

Valladas H *et al* 1987 Thermoluminescence dates for the Neanderthal burial site at Kebara cave in Israel *Nature* **330** 159–60

Valladas H *et al* 2008 Radiometric dates for the Middle Palaeolithic sequence of Payre (Ardèche, France) *Quat. Geochronol.* **3** 377–89

Voinchet P, Bahain J J, Falgueres C, Laurent M, Dolo J M, Despriee J, Gageonner R and Chausse C 2004 ESR dating of quartz extracted from Quaternary sediments; Application to fluvial terraces system of northern France *Quaternaire* **15** 135–41

Voinchet P, Despriée J, Tissoux H, Falguères C, Bahain J J, Gageonnet R, Dépont J and Dolo J M 2010 ESR chronology of alluvial deposits and first human settlements of the Middle Loire Basin (Region Centre, France) *Quat. Geochronol.* **5** 381–4

Yokoyama Y, Falguères C and Quaegebeur J P 1985 ESR dating of quartz from quaternary sediments: first attempt *Nucl. Tracks Radiat. Meas.* **10** 921–8

IOP Publishing

Spectroscopic and Microscopy Techniques for Archaeological and Cultural Heritage Research (Second Edition)

Ashutosh Kumar Shukla

Chapter 6

The development of low frequency EPR spectroscopy for studying objects with cultural heritage significance

Joseph P Hornak

Conventional, high frequency electron paramagnetic resonance (EPR) spectroscopy operating at 9 GHz has proven to be a useful tool for studying objects with cultural heritage significance. Unfortunately, electron paramagnetic resonance spectroscopy is invasive as it is limited to objects a few mm in size. This chapter describes a non-destructive form of EPR spectroscopy operating at 100–500 MHz that can examine larger objects. We refer to this as low frequency EPR spectroscopy. Details of the instrument and cultural heritage applications are presented.

6.1 Background

Museum curators, art historians, restoration scientists, and archaeologists have many analytical tools at their disposal to study objects with cultural heritage significance. These analytical tools include:

1. ultraviolet–visible (UV–vis) absorption, reflection, fluorescence, and phosphorescence spectroscopy [1–5];
2. infrared (IR) (also referred to by the instrumentation type Fourier transform IR (FTIR)) absorption and reflection spectroscopy [6–11] in the near, shortwave, mid, and far regions of the spectrum;
3. Raman spectroscopy (RS) [12], surface enhanced Raman spectroscopy (SERS) [13, 14], and surface enhanced resonance Raman spectroscopy (SERRS) [15];
4. x-ray powder diffraction (XRPD) [16];

5. x-ray fluorescence (XRF) [17, 18], energy dispersive XRF (EDXRF), and wavelength dispersive XRF (WDXRF), and confocal XRF [19] spectroscopy;
6. mass spectroscopy (MS), laser photolysis gas chromatography mass spectroscopy (LP GC–MS) [20, 21], and pyrolysis gas chromatography mass spectroscopy (Py GC–MS) [22];
7. neutron activation analysis (NAA) [23];
8. particle-induced x-ray emission (PIXE) spectroscopy [24, 25];
9. Mössbauer spectroscopy [26];
10. conventional liquid phase nuclear magnetic resonance (NMR) [27, 28], solid state NMR (SSNMR) [29, 30], and unilateral time resolved NMR spectroscopy [31, 32]; and
11. electron paramagnetic resonance (EPR) spectroscopy.

With so many techniques to choose from, the choice can be daunting to the curator, art historian, restoration scientist, and archaeologist. These individuals should keep in mind that no one of these analytical techniques can analyze all varieties of cultural heritage objects. Instead, combinations are used to portray the complete picture of an object. This chapter addresses a new development in EPR spectroscopy that makes it non-invasive and portable, and thus a more viable tool for analyzing objects with cultural heritage significance.

When analyzing cultural heritage objects, it is important for the technique to be non-invasive and non-destructive. These two terms have slightly different meanings in the cultural heritage and analytical chemistry communities. To the analytical chemist a non-destructive analytical technique is one that does not destroy the analyte rendering it useful for analysis by another technique or at a later date by a new technique. A non-invasive technique does not alter the original object. From the perspective of the owner of a cultural heritage object, any invasive analytical technique might be considered destructive, or at least minimally destructive, as sampling an object, no matter how small, will destroy its integrity. UV–vis spectroscopy is considered non-invasive and non-destructive, while conventional EPR and NMR are considered invasive and destructive.

A second factor to consider is the depth of analysis. Does the technique analyze the surface or the bulk? A bulk analytical technique can reveal information hidden by a surface layer that is invisible to analysis by a surface technique. Portability of the analytical device means the difference between bringing the object to the analytical instrument or being able to analyze the object on site. Perhaps the most important feature of an analytical technique to the analyst is if the result of the analysis is it conclusive, or inconclusive and requires further analysis by another technique.

Conventional 9 GHz electron paramagnetic resonance (EPR) spectroscopy, referred to as high frequency EPR, is a bulk analysis and has been used to study many items with cultural heritage significance [33] including pigments [34–38], ceramics [39–41], painting varnishes [42], marble and limestone objects [43–45], paper [46], wall paintings [47], and rock paintings [48]. Until recently [49, 50], EPR was considered invasive to all but planar 1 cm × 2 cm objects and not portable. This chapter summarizes the work from the Rochester Institute of Technology (RIT) Magnetic Resonance Laboratory (MRL) developing non-invasive and portable EPR spectroscopy for studying objects with cultural heritage significance such as paintings, ceramics, and marble objects. The spectrometer operates in the 100–500 MHz radio frequency (RF) region of the electromagnetic spectrum, instead of the microwave region. This form of EPR spectroscopy is therefore referred to as low frequency electron paramagnetic resonance (LFEPR) spectroscopy.

Although LFEPR spectroscopy has a greater limit of detection than conventional high frequency EPR spectroscopy, LFEPR is completely non-invasive as it can look at small surface regions of any size larger object. Like high frequency, it is also non-destructive. LFEPR is limited to paramagnetic materials with peak-to-peak line-widths (Γ_{PP}) less than ~25 to 30 mT. The materials found in many objects with cultural heritage significance satisfy this requirement, however many other pigments and clays contain ferro/ferrimagnetic iron. These objects are not paramagnetic and therefore lack a conventional EPR signal. They do however possess a low field microwave absorption signal [51] on the LFEPR spectrometer which can be utilized to characterize the object. In the LFEPR spectrometer this will be referred to as the low field RF absorption signal.

In 1990, the RIT MRL constructed an LFEPR spectrometer for studying *in vivo* free radicals [52]. Although the work resulted in a few publications about the development of the instrument and an imaging approach based on the phase of the modulation frequency [53], the intended application was never realized by them but was by others [54, 55]. The LFEPR was used occasionally over the next two decades for unique projects such as determining the paramagnetic metal and free radical content in sands. In late 2010 Dr Nicholas Zumbulyadis, a retired senior scientist from Eastman Kodak and porcelain expert [56–58], proposed the idea to the RIT MRL of using LFEPR to study ceramic objects with cultural heritage significance. The spectrometer had a large bore electromagnet that could potentially accommodate objects as large as 30 cm in diameter and consequently non-destructively and non-invasively record their EPR spectra.

Several spectrometer modifications would be needed to achieve the goal of looking at ceramic objects. The original system had a narrow sweep width of 6 mT, which needed to be increased to observe the broad spectra from paramagnetic transition metal ions. A spectrometer for studying ceramics would need to scan from zero to more than 20 mT. The spectrometer would need some way to accommodate objects larger than would fit inside the approximately one cm diameter sample probe. Surface coil technology was proposed which in turn required large volume modulation coils.

Since its initial modification, the spectrometer design for investigating cultural heritage objects has evolved [49, 50]. Details of the current spectrometer design are presented in enough detail and with references to allow construction of a similar unit by others interested in doing similar studies. This chapter contains applications of the spectroscopy to several standards and test samples that demonstrate the capability of the technique under controlled situations.

6.2 Theory

6.2.1 Electron paramagnetic resonance

Paramagnetic materials possess a magnetic moment, typically from unpaired electrons but also from electron orbitals. At room temperature and in the absence of a magnetic field, the magnetic moments of paramagnetic materials are randomly oriented. In an externally applied magnetic field, these magnetic moments can be aligned either with or opposed to the magnetic field. At a temperature T, the ratio of the number of oppositely (N^+) to aligned (N^-) magnetic moments is given by Boltzmann's equation

$$\frac{N^+}{N^-} = e^{-(\Delta E/kT)}, \qquad (6.1)$$

where ΔE is the energy difference between the upper (+) and lower (–) states, and k is Boltzmann's constant 1.381×10^{-23} J K^{-1}. These magnetic moments, be they spin or orbital, are the focus of EPR spectroscopy.

Continuous wave (CW) EPR is performed by observing the absorption of photons at a fixed Larmor frequency (ν) as a function of an applied magnetic field (B_o) and satisfying the general resonance condition

$$h\nu = g\beta B_o, \qquad (6.2)$$

where h is Planck's constant 6.626×10^{-34} J s, β the Bohr magneton 9.274×10^{-24} J T^{-1}, and g the Landé g-factor [59]. LFEPR spectrometers operate in the approximate range $100 \leqslant \nu \leqslant 500$ MHz, while the majority of conventional EPR spectrometers operate at $\nu \geqslant 9$ GHz. The Larmor frequency is applied to the sample in the form of an oscillating magnetic field in a sample probe or cavity resonator, also resonant at ν and critically coupled to the impedance of the spectrometer bridge. The swept magnetic field is typically modulated at an audio frequency $10 \leqslant f_{\text{mod}} \leqslant 100$ kHz and the signal is phase sensitive detected at f_{mod}, thus producing a first derivative absorption peak signal. Figure 6.1 presents the relationship between the magnetic field, the energy levels, the Larmor frequency, and the absorption and first derivative signal.

All acquisition parameters being constant, the area under a spectral absorption peak is proportional to the sample concentration. The area increases as the peak-to-peak signal S_{PP} and linewidth Γ_{PP} of the first derivative of the absorption peak. See figure 6.1 for a pictorial definition of S_{PP} and Γ_{PP}. The signal-to-noise ratio (SNR) of the spectrum of a standard such as 2,2-diphenyl-1-picrylhydrazyl (DPPH) is a measure of the quality of a spectrum from an EPR spectrometer. The signal is taken as S_{PP} and the noise as the standard deviation of the baseline.

Most EPR and LFEPR spectrometers have an automatic frequency control (AFC) which keeps the frequency ν of the source locked to the frequency of the resonant sample probe. The coupling of the probe to the spectrometer is typically not controlled during the recording of a spectrum because unless the temperature changes, the coupling is fairly constant for paramagnetic materials.

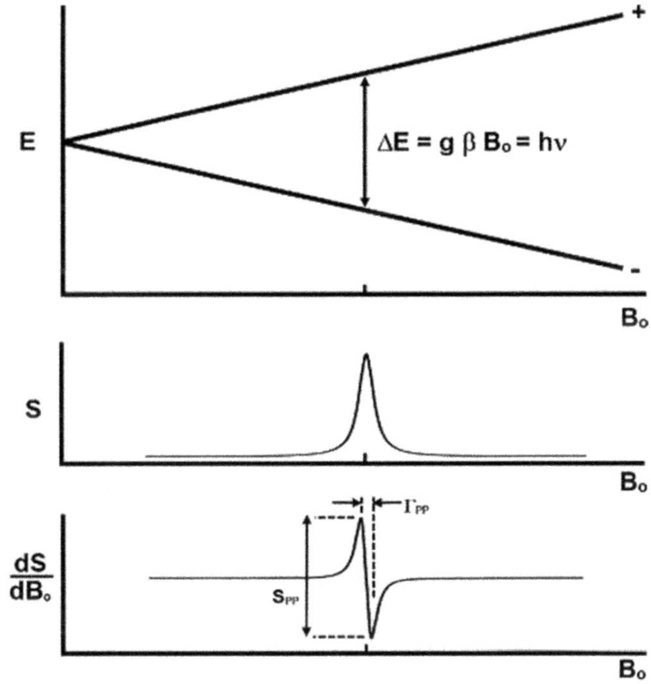

Figure 6.1. Energy level diagram, absorption spectrum, and first derivative spectrum for a spin 1/2 system. ΔE is the energy difference between the two spin states at frequency ν. The first derivative of the absorption has peak-to-peak signal S_{PP} and linewidth Γ_{PP}.

For some paramagnetic species, *g* is isotropic, but for others *g* is anisotropic and a *g* tensor is needed to describe *g*. Adopting a coordinate system for the paramagnetic species where *Z* is the principle axis of the crystal or molecule and g tensor, this anisotropy can be reported in terms of a g_X, g_Y, and g_Z. Often the principle axis is unknown and a g_1, g_2, and g_3 are reported instead. In some species the geometry is such that a *g*-factor parallel ($g_{//}$) and perpendicular ($g_{\perp}$) to the B_o will suffice. In systems where the orientations of paramagnetic species are fixed, such as in solids, asymmetric spectral absorption line shapes result.

Paramagnetic species with nuclear spins experience hyperfine interactions between the electron and nuclear spins, resulting in splitting of the energy levels and hence multiple spectral absorption lines. Similarly, spin–spin electron interactions can be seen in paramagnetic species where the electron spin is greater than 1/2. Both of these interactions have an orientation dependence contributing to asymmetric spectral lines. Unlike the energy level diagram of figure 6.1 for a single spin 1/2 electron, the energy level diagram for systems with electron–nuclear hyperfine and or electron–electron spin–spin interactions possess split energy levels even at zero applied magnetic field. This is referred to as the zero-field splitting. This results in predictably different low and high frequency EPR spectrum. The Mn(II)

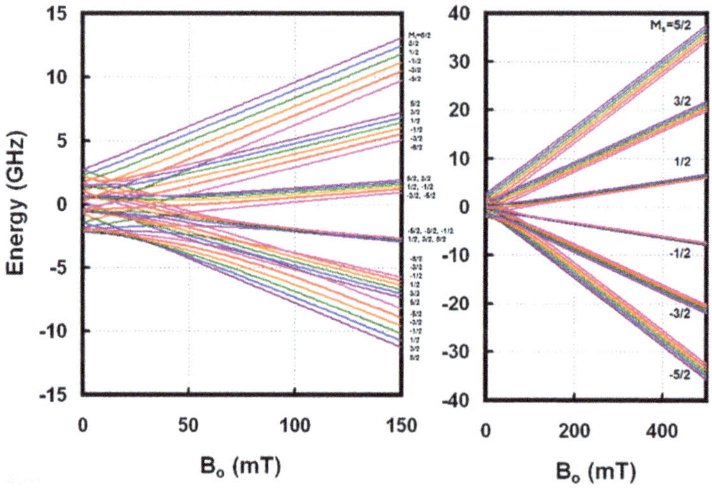

Figure 6.2. Energy levels of the Mn(II) ion as a function B_o. The high field plot labels the six M_S electron spin manifolds and the low field plot has the M_I nuclear spin states labeled.

ion with a nuclear spin of $I = 5/2$ with $M_I = 5/2, 3/2, 1/2, -1/2, -3/2$, and $-5/2$; an electron spin of $S = 5/2$ with $M_S = 5/2, 3/2, 1/2, -1/2, -3/2$, and $-5/2$; and energy level diagram presented in figure 6.2 is an example of such a system [60].

One last influence on the spectral line shape is spin relaxation. Equation (6.1) presents the equilibrium distribution of spins between two energy levels. If a system of spins is perturbed from this distribution it will return to the equilibrium configuration with a time constant called the spin-lattice relaxation time (T_1). The name implies that the energy a spin possesses in the excited state is dissipated to the lattice in the form of vibrational energy, or in the case of other non-solid states of matter, the surrounding molecules in the form of vibrational, rotational, and translational energy. Energy may also be transferred to other spins which does not change N^+/N^- but does cause of loss of phase coherence of the spins. This process is called spin–spin relaxation and is characterized by a spin–spin relaxation time (T_2). T_1 and T_2 affect the spectral linewidth.

$$\Gamma_{PP} \propto \frac{1}{T_1} + \frac{1}{T_2}. \tag{6.3}$$

Lowering the sample temperature increases the relaxation times and hence decreases the linewidth. Exciting too many spins without giving them a chance to relax to the ground state can result in a saturated spin system and a diminished signal. Having too many spins in close proximity to each other results in a very short T_2 value and what is referred to as an exchange broadened spectral line.

In general, paramagnetic materials have similar, or at least predictably different, spectra on high and low frequency EPR spectrometers. The common solid EPR standard DPPH has the same g-factor at all Larmor frequencies [59].

6.2.2 Ferro/ferrimagnetic materials in LFEPR

Ferromagnetic materials have the magnetic moments within small regions called domains aligned in one direction. Ferrimagnetic materials are similar but not all the magnetic moments are aligned, some point in the opposite direction. Paramagnetic, ferromagnetic, and ferrimagnetic materials possess a spin resonance signal that follow the resonance equation (equation (6.2)). As the frequency of the spectrometer decreases, the absorption decreases proportionately in B_o. In paramagnetic materials this signal can be followed to zero field. In ferromagnetic and ferrimagnetic materials the presence of another signal at low B_o values overwhelms the ferromagnetic resonance signal making it undetectable.

To understand the nature of this other signal, consider the relationship between the amount of magnetic field produced within a sample (B) as a function of the applied magnetic field intensity (H). B is related to H through the permeability (μ) of the sample.

$$B = \mu H. \tag{6.4}$$

In a specific diamagnetic or paramagnetic material μ is a constant with respect to H and very close in value to the permeability in a vacuum (μ_o) $4\pi \times 10^{-7}$ H m^{-1}. The same is not true for ferromagnetic and ferrimagnetic materials where

$$B = \mu_o(H + M), \tag{6.5}$$

and M is the magnetization induced by H inside the sample. For EPR where samples are paramagnetic, and NMR where samples are generally diamagnetic, the amount of magnetic field produced in the sample is proportional to the amount of magnetic field applied to the sample. For this reason, the EPR, NMR, and magnetic resonance imaging (MRI) communities tend to use B_o for the applied magnetic field. In ferromagnetic resonance (FMR) and electron magnetic resonance (EMR), where samples are ferromagnetic or ferrimagnetic, the distinction between H and B is important. Unfortunately, this usage can be a source of confusion when comparing results from the EPR, FMR, and EMR communities. The following symbolism is adopted to minimize the confusion. An applied magnetic field H in air will be reported as $\mu_o H$ with units of Tesla making it easy to compare to B_o used in EPR.

Ferromagnetic and ferrimagnetic materials retain a history of the magnetic field they were exposed to and μ is not constant with respect to H. If an unmagnetized sample at $H = 0$ is exposed to an increasing H, B increases nonlinearly to a value called the saturation magnetization (B_S). Beyond B_S, B increases linearly with respect to H. Decreasing H to zero after reaching B_S does not send B to zero. Instead, the sample retains magnetization known as the remanent magnetization (B_R). H must be decreased to a negative value called the coercivity (H_C) in order for the B to equal zero. Decreasing H further produces a $-B_S$ and increasing H from this point causes B to follow the counterclockwise path around the hysteresis curve in the direction of the arrows.

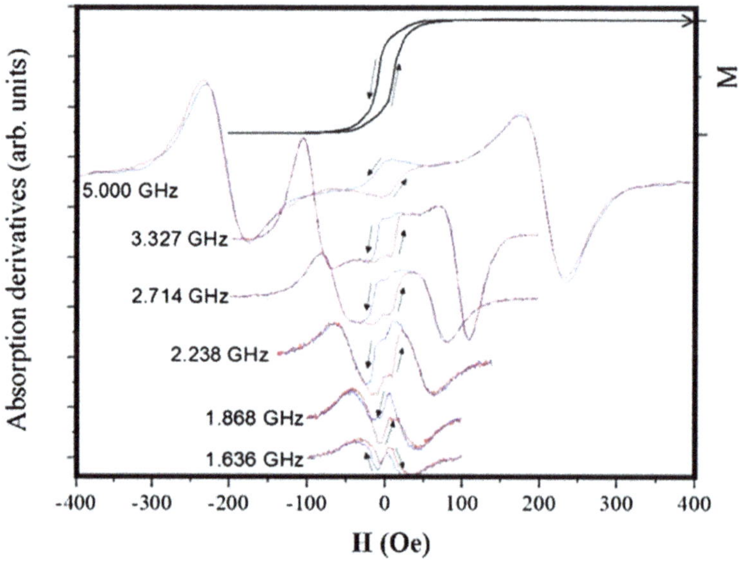

Figure 6.3. For an arbitrary ferromagnetic material, an *M–H* plot (top), and first derivative FMR spectra scanned upfield and downfield at six different frequencies (bottom). Reproduced from [51], with the permission of AIP Publishing.

There are two types of magnetic hysteresis curves which are often confused. The *M–H* curve is horizontal at the saturation magnetization and the $M = 0$ crossing of the *H*-axis is the intrinsic coercivity (H_{Ci}), which is different than H_C. An *M–H* curve is depicted at the top of figure 6.3.

Consider the FMR spectrum for an arbitrary ferrimagnetic material with a $g \approx 1.85$ and $\Gamma_{PP} \approx 60$ mT. Figure 6.3 displays the hypothetical spectrum at different frequencies for B_o scans from −400 mT to +400 mT and then back to −400 mT. At 5 GHz the $g \approx 1.85$ absorption is clearly visible. Between $B_o = \pm B_S$ the spectrum shows a hysteresis behavior known as low-field microwave absorption. As ν is decreased, the B_o position and S_{PP} of the $g \approx 1.85$ peak decrease. At ν values less than approximately $g\beta\ B_S/h$, S_{PP} is small compared to the low-field microwave absorption signal.

An LFEPR spectrometer with a field scan between $0 < B_o < 18$ mT operates at $\mu_o H$ values $<B_S$, and also less than $\sim 2\mu_o H_C$. Repeated upfield followed by downfield scans of B_o between zero and 18 mT will trace out mini hysteresis curves between 0 and 18 mT, but never exceeding the right/left full hysteresis curve limits. See figure 6.4 for a possible mini hysteresis curve. A mini hysteresis curve starting at $M = 0$ and $H = 0$ increases in *M* until the maximum *H* of the scan is reached. We refer to this point as (H', M') in figure 6.4. The curve then decreased in *M* until $H = 0$ is reached. Subsequent loops start from this point (H', M'), increase to the same (H', M') value then decrease to the M' value at the end of the first loop. The difference between the LFEPR signal present after the first loop and the signal at the start for the unmagnetized sample is related to the remanent magnetization present after the first

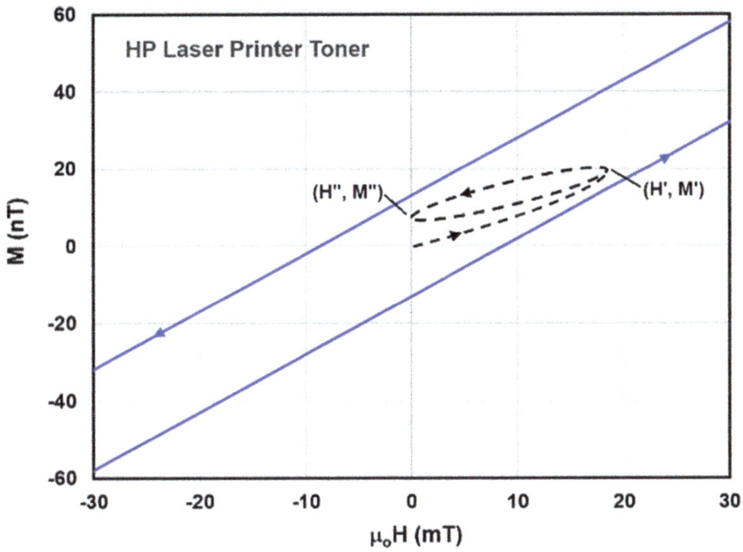

Figure 6.4. The center portion of the M–H magnetic hysteresis loop (solid line) for Fe_3O_4 nano particles. Dashed line indicates a possible mini hysteresis loop for cycling a sample between 0 and 18 mT.

cycle ($M''-M'$). Just like on the full hysteresis curve, μ at any point around the mini hysteresis curve will also be B/H.

The μ of a ferromagnetic or ferrimagnetic sample in the single turn solenoid sample probe influences the resonance frequency as well as the coupling of the sample probe to the LFEPR spectrometer. Increasing B_o increases μ. The AFC compensates for changes in the resonant frequency of the sample probe due to changes in μ, but not changes in the coupling. Assuming the probe is perfectly coupled at the start of a scan, more power is reflected from the sample probe as B_o increases and μ changes. Modulating B_o at f_{mod} also modulates the reflected power at frequency f_{mod} thus creating a signal in the phase sensitive detector at f_{mod} proportional to $\Delta\mu/\Delta H$. The greater the modulation amplitude at f_{mod} the greater the ΔH and signal.

Figure 6.5 presents the signal from polymer coated, 50–150 nm ferrimagnetic Fe_3O_4 particles on the LFEPR spectrometer when the applied magnetic field is swept up and down between 0 and 18 mT. This material has a clearly visible $g \approx 1.85$, $\Gamma_{PP} \approx 110$ mT peak on a 9.1 GHz FMR spectrometer [61], but at 419 MHz all that is seen is the response of the spectrometer to the relative change in μ as a function of H.

The smaller H_C is compared to H' the more a mini hysteresis curve is forced to follow the major hysteresis curve, and the larger the change in $\Delta\mu/\Delta H$, and the larger the area of the LFEPR signal hysteresis. The larger H_C is compared to H' the opposite is true and the smaller the area of the LFEPR signal hysteresis. Therefore, in LFEPR, ferromagnetic and ferrimagnetic pigments and clays cannot be identified by g and Γ_{PP}, but may be characterized by their $\Delta\mu/\Delta H$ effect in a field cycled

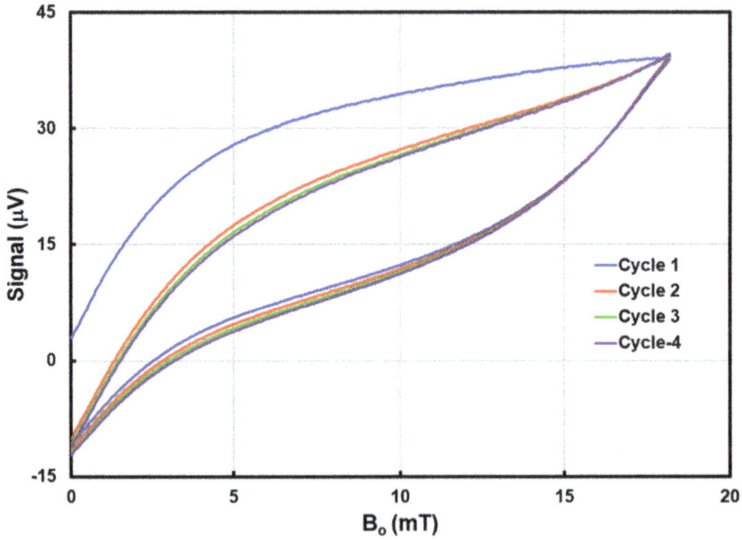

Figure 6.5. The signal from polymer coated, 50–150 nm ferrimagnetic Fe_3O_4 particles on an LFEPR spectrometer. B_o was cycled between 0, 18.2, and 0 mT in 2000 even steps four times with a dwell time of 215 ms/point, 0.03 s time constant, and 94 μT modulation amplitude at 10 kHz.

spectrum. Currently, ferromagnetic and ferrimagnetic materials are indistinguishable with the LFEPR spectrometer. For this reason they will be referred to as ferro/ferrimagnetic materials.

6.3 Instrumentation

The continuous wave LFEPR spectrometer is described in enough detail to permit an understanding of its operation and replication, as currently there are no commercial LFEPR spectrometers operating below 1 GHz. Additional details can be found in the original literature [49, 50, 52]. The RIT wide-line LFEPR spectrometer has two basic configurations distinguishable primarily by the magnet. The first is a benchtop unit with a 30 cm diameter solenoidal magnet and second is a smaller, more mobile unilateral unit referred to as the EPR mobile universal surface explorer (MOUSE). Both of these configurations utilize a common EPR radio frequency (RF) source and bridge, magnetic field modulation amplifiers, signal lock-in amplifier, and computer control hardware and software. The magnetic field modulation coils and probes are different for the benchtop and MOUSE units. Their arrangement is depicted in the basic block diagram of a CW LFEPR presented in figure 6.6. The common components are described first.

The spectrometer employs homodyne signal detection and incorporates an automatic frequency control (AFC) that keeps the frequency of the source (ν) matched to the resonant frequency of the sample probe (ν_P). Frequency ν is sent into the sample probe where it can be absorbed by the sample when resonance occurs. The magnetic field is modulated at f_{mod} as it is scanned from $B_{o\text{Min}}$ to $B_{o\text{Max}}$. When

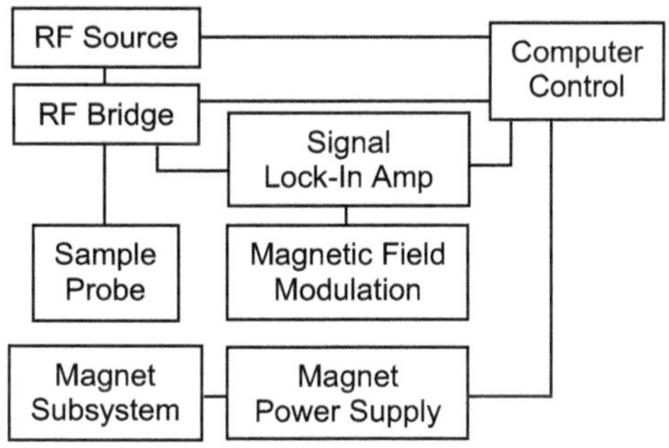

Figure 6.6. Basic block diagram of a CW LFEPR spectrometer.

resonance occurs, the signal is modulated at f_{mod} yielding frequency $(\nu + f_{\mathrm{mod}})$ coming back from the sample probe. Within the RF bridge this frequency is mixed with ν in a doubly balanced mixer to produce $(2\nu + f_{\mathrm{mod}})$ and f_{mod}, the larger of which is removed by a low pass filter. The small frequency component is sent to the signal lock-in amplifier where it is phase sensitive detected against the f_{mod} reference to yield the signal. In this spectrometer, ν is adjustable between 100 and 500 MHz and f_{mod} is 10 kHz.

An automatic frequency control (AFC) is a vital part of a continuous wave electron paramagnetic resonance (EPR) spectrometer. As such, it warrants a more detailed description. An AFC is necessary because changes in temperature or the sample can alter the resonant frequency of the sample probe. Without the AFC to keep the two frequencies the same, the EPR signal will be degraded. Several approaches to implementing an AFC on an EPR spectrometer have appeared in the literature [59, 62–66]. Most of these rely on the generation of an electrical voltage (V_E) proportional to the difference between ν_{o} and ν_P, which is used to change ν to equal ν_P. Most AFCs perform this task with analog circuitry continuously throughout a scan. We present a hybrid design that generates V_E via traditional analog circuitry but uses software code to determine if ν_{o} needs to be changed and if necessary, changes it by a user definable amount before the acquisition of each data point. The LFEPR has ten possible sample probes and therefore the AFC needed to function with all probes.

Figure 6.7 presents the amount of power and its phase reflected from an LC circuit such as an EPR sample probe as a function of frequency. When $\nu = \nu_P$ the reflected power is minimal and phase is zero. At frequencies less than and greater than ν_P, the reflected power increases. The phase (ϕ) becomes less than zero in one direction and greater than zero in the other. If frequency sidebands are imposed on ν at $\pm f_{\mathrm{AFC}}$, and these three frequencies are directed at the sample probe when $\nu = \nu_P$, there should be no reflected power from the sample probe at f_{AFC} relative to ν as the $(\nu + f_{\mathrm{AFC}})$ and $(\nu - f_{\mathrm{AFC}})$ components cancel. However, if $\nu_P > \nu$ or $\nu_P < \nu$, the $\nu + f_{AFC}$ and $\nu - f_{\mathrm{AFC}}$ components do not cancel and in one case the phase of the reflected f_{AFC} is positive and the other negative. The size of f_{AFC} depends on $(\nu_P - f_{\mathrm{AFC}})$, the Q of the sample probe, and the difference between ν and ν_P.

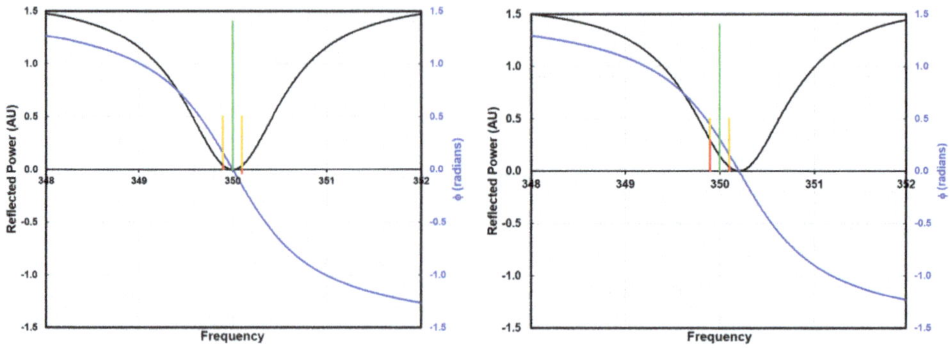

Figure 6.7. Plots of the reflected power (black) and its phase (blue) as a function of frequency. Superimposed on the plots are ν_o (green), $\nu \pm f_{AFC}$ (orange), and the reflected values of $\nu \pm f_{AFC}$, (red) for (left) $\nu = \nu_P$ and (right) $\nu_o < \nu_P$.

$$A_1 e^{i(2\pi f_{AFC}t+\phi_1)} + A_2 e^{i(2\pi f_{AFC}t+\phi_2)} = e^{i(2\pi f_{AFC}t)}[A_1 e^{i\phi_1} + A_2 e^{i\phi_2}]. \tag{6.6}$$

The signal from the sample probe is mixed with ν to produce 2ν, $(2\nu + f_{AFC})$, $(2\nu - f_{AFC})$, f_{AFC}, $-f_{AFC}$, and DC. A low pass filter removes 2ν, $(2\nu + f_{AFC})$, $(2\nu - f_{AFC})$ leaving f_{AFC}, $-f_{AFC}$, and DC. The sum of f_{AFC} and $-f_{AFC}$ will be the same frequency with a phase dependent on the amplitudes and phases of the positive and negative frequency components. These three components are sent to a lock-in amplifier for phase sensitive detection at f_{AFC}. The output of the lock-in amp is voltage V_E. It should be noted that when $\nu = \nu_P$ the output of the AFC is zero. For the two cases when $\nu \neq \nu_P$ one yields a positive V_E and the other a negative V_E. V_E is digitized and used by LabVIEW code to change the source frequency.

Since the phase change at resonance as a function of frequency can go from positive to negative or negative to positive, the direction of the frequency change calculated by the LabVIEW code for $+V_E$ and $-V_E$ must be determined empirically. A quick scan of a spectrum can determine this.

6.3.1 RF source and bridge

The radio frequency bridge consists of three parts: one is for sample excitation and signal detection, a second is for automatic frequency control (AFC), and the third is for sample probe tuning. The three parts and their components are presented in figure 6.8. All bridge components are specified to operate between at least 100 and 500 MHz and are interconnected with rigid 50 Ω rigid coaxial cable with SMA connectors.

RF energy at frequency ν is provided by a 0–2 GHz frequency synthesizer (Sanford Research Systems SG-382). Output from the synthesizer is divided into four equal parts in the bridge by a power divider and sent into the four branches of the bridge. One branch is used for sample excitation, a second is for the AFC, a third is a reference for the AFC detection and marker for tuning, and the fourth is a reference for the signal detection.

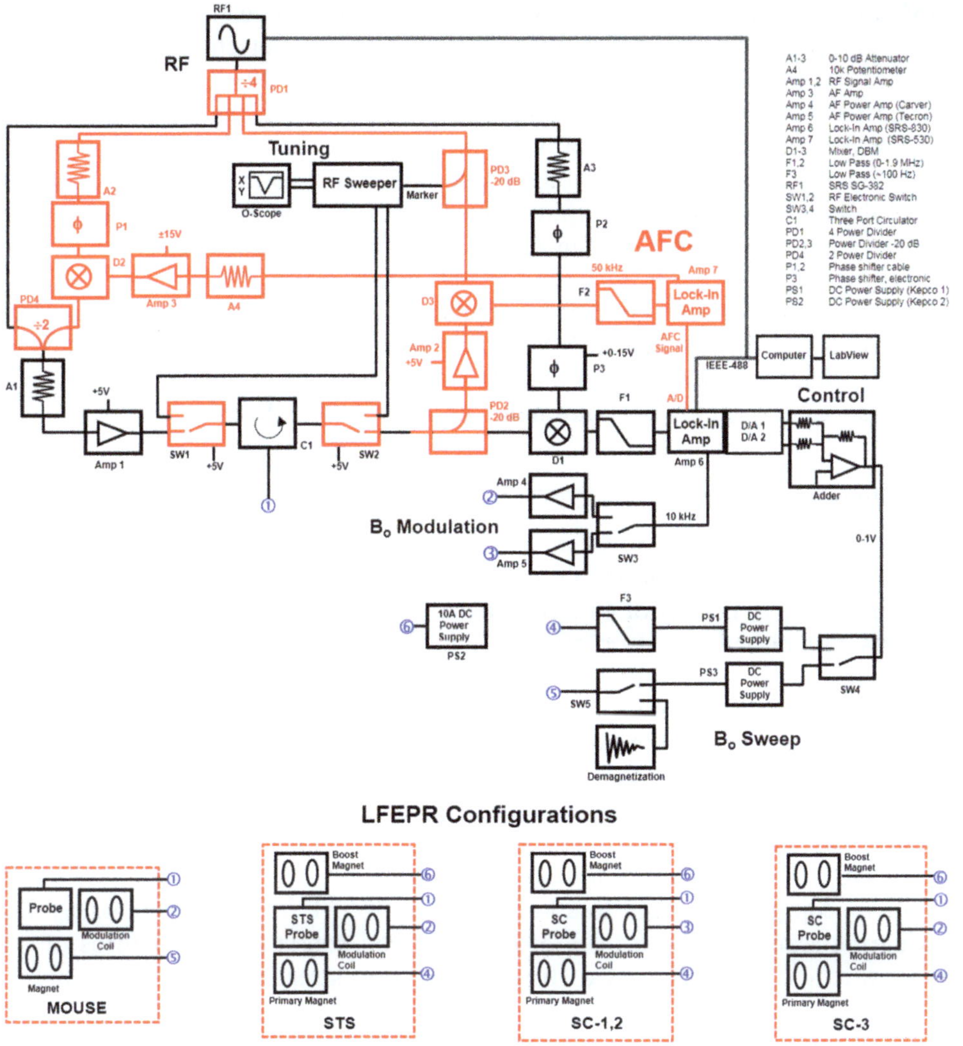

Figure 6.8. A schematic block diagram of the LFEPR spectrometer.

The sample excitation power is sent to the sample coil via a 0–10 dB attenuator, small power amplifier, and three-port circulator. The AFC branch imposes two frequency sidebands at ±100 kHz on ν utilizing a doubly balanced mixer (DBM) and a high to low impedance driving amplifier powered by a signal from a lock-in amplifier (Amp 7) (Sanford Research Systems, SRS-530). The output power of the AFC branch is adjustable via a 0–10 dB attenuator. The phase of the AFC branch is matched to that from the sample excitation branch via a delay line before the two are combined. Signals coming back from the sample coil are sent to a −20 dB directional coupler. The −20 dB component is sent to a signal amplifier and then to a DBM where it is combined with RF power from the AFC reference arm. The output of this

DBM is sent through a low pass filter to remove the sum frequencies ($2\nu \pm 100$ kHz) and retain the difference frequencies (±100 kHz) for detection by the lock-in amplifier (Amp 7). Power from the other arm of the −20 dB power divider is directed to the signal DBM. The reference is derived from the fourth arm of the bridge which allows attenuates via A3 to properly bias the DBM. The phase of this reference is adjusted via an electronic phase shifter (P2) so the signal possesses the proper absorption line shape. Output of this DBM is filtered to retain the signal at f_m which is sent to the signal lock-in amplifier (Amp 6) (Sanford Research Systems SRS-830).

On either side of the three-port circulators are two single throw, two pole electronic RF switches (SW1, SW2) that switch in and out an RF sweeper (Wavetek 7062) for tuning the probe. Tuning consists of observing the resonance absorption on an oscilloscope and matching it to 50 Ω and setting ν to ν_P.

6.3.2 Magnet subsystem

The magnet of the benchtop system is a 800 turn 30 cm inside diameter (ID), 45 cm length solenoid wound with #10 copper wire. The end 3 cm of each opening of the solenoid had a turn density of 40 turns/cm while the center 39 cm value was 14 turns/cm. This diminished the drop-off in B_o towards the ends of the solenoid. An additional set of ten windings were placed at various locations along the length the solenoid to help homogenize B_o. The total series resistance from all the windings was 3.8 ohms. This magnet produced a maximum of 18.7 mT of magnetic field. A 400 μF capacitor was placed in parallel across the windings of the coil to act as a low pass filter, removing the modulation frequency induced in the magnet windings by the larger diameter modulation coils. The solenoidal magnet was driven by a 0–36 V. 0–30 A DC power supply which was controlled by a 0–1 V DC signal. A 15 cm ID Helmholtz coil pair was used to add or subtract 16 mT from the scanned B_o, thus making the scan range −16 to +35 mT. This coil consisted of 200 turns of #10 copper wire and was driven by a 10 A current from a second DC power supply.

6.3.3 B_o modulation subsystem

The B_o modulation subsystem consists of a modulation frequency source, a power amplifier, and an array of possible modulation coils. The SRS-830 lock-in provides f_m and allows the user to set its frequency, amplitude, and phase. The power amplifier takes the modulation frequency and creates a sufficient amount of power to drive the modulation coils. Two power supplies were utilized depending on the chosen modulation coil. The smaller of the two was a Carver TFM-15 audio frequency amplifier capable of driving an approximately 100 W of power into an 8 Ω load. The second was a Tecron 7570 capable of driving a wide range of load impedances and is specked at 1000 W into a 4 Ω load. Although both amplifiers had dials that can adjust the output power given a fixed input, they were not used as they altered the phase of the modulation current. Instead the power output was varied by changing the input signal.

Table 6.1. Magnetic field modulation subsystems.

Name	STS probe	10 cm Solenoid	15 cm Helmholtz	3 mm Surface coil	MOUSE
Coil type	Saddle coil	Solenoid	Helmholtz	Figure-8	Yoke
Resistance (Ω)	12.6	2.3	1.8	5.5	9.2
ID (cm)	3.8	10	15	Two 2 × 1.5	Two 4 × 1
Amplifier	Carver	Techron	Techron	Carver	Carver
Max B_{Mod} (mT)	0.65	2.06	0.22	2.48	1.99

The carver amplifier was used to power a pair of saddle coils around the shield of the STS coils, a *figure-8* coil available for use with the 3 mm surface coil, and the MOUSE modulation coils. The STS saddle coil consisted of 136 turns of #32 copper wire wound in a rectangular form on a 4.3 cm diameter. The saddle coil for the STS produced a maximum modulation field (B_{Mod}) of 0.65 mT while the figure 6.8 produced 2.48 mT. The modulation field from the MOUSE will be discussed in the EPR MOUSE section.

The Tecron amplifier drove either a 15 cm ID, 200 turns of #10 copper wire Helmholtz coil or a 155 turn solenoidal coil of #14 square copper wire. These larger modulation coils were used with the 2 cm surface coil or the 3 cm surface coil when the figure eight coil was not utilized. The 10 cm ID coil was capable of delivering a 2.06 mT modulation field, while the 15 cm ID coil could deliver a 0.22 mT field. The characteristics and capabilities of all the modulation coils are summarized in table 6.1.

6.3.4 Sample probes

6.3.4.1 Single turn solenoid

RF sample probes for use with the LFEPR solenoidal magnet are either volume or surface coil (SC) style. The volume coils are single turn solenoid (STS) style coils which detect an EPR signal from samples placed within them. The surface coils are placed on a sample and detect a signal from the sample adjacent to the coil. The size, resonance frequency, relative sensitivity, and Q of the coils are summarized in table 6.2.

The STS coils can accommodate a 15.8 mm diameter by 3.2 cm length sample with a volume of 6.3 cm^3. The inductor of the coil is constructed of 80 μm thick copper foil in the form of a cylinder, 2 cm diameter and 3 cm length. The copper cylinder had a longitudinal gap across which capacitors were placed and five equatorial slits to allow the modulation frequency of the magnetic field to better pass through to the sample. All STS coils had the same physical dimensions and inductance (L). The capacitance (C) differed to give the variety of resonance frequencies according to

$$\nu = \frac{1}{2\pi\sqrt{LC}}. \tag{6.7}$$

Table 6.2. Properties of the different sample probes.

Type	Size	Inductor	Support	ν (MHz)	Q	Relative sensitivity
Surface coil	$d = 2.9$ mm	5 Turn coil	PE	245	75	1.00
	$d = 2$ cm	Spiral	PTFE circuit board	271	109	0.09
MOUSE surface coil	$d = 2.9$ mm	5 Turn coil	PE	355	100	
MOUSE STS	$d = 2.9$ mm	Ag foil	PTFE	421	65	
STS	$V = 6.3$ cm^3	Slotted Cu foil	polyvinyl chloride	150	127	0.01
				205	196	0.02
				251	212	0.02
				299	186	0.03
				354	180	0.01
				400	444	0.03
				440	445	0.02

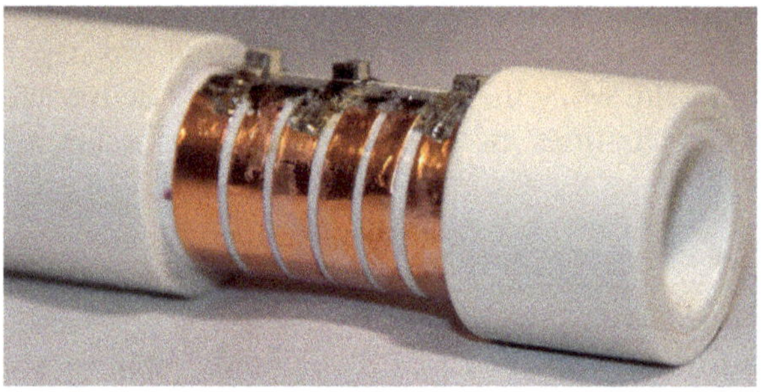

Figure 6.9. The slotted STS inductor with chip capacitors on the polyvinyl chloride support.

All the copper inductors were supported by a polyvinyl chloride tube (figure 6.9) and surrounded by a 38 mm ID, 76 mm length, 0.5 mm thick cylindrical silver-plated brass shield. (See figure 6.10.) RF energy was coupled into and out of the STS via a one-turn, 3.7 cm ID loop of #18 copper wire. The STS coupling loop was connected to the spectrometer via a length of 50 Ω coaxial cable.

6.3.4.2 Surface coils

There were two surface soils, a 2.9 mm ID and a 2 cm OD Spiral. Both of which were inductively coupled to the 50 Ω cable to the spectrometer bridge. Coupling was varied by changing the distance between the coupling loop and the surface coil.

The 2.9 mm diameter coil consisted of five turns of # 26 copper wire wound along a 2.9 mm OD polyethylene cylinder. A 5 pF non-magnetic chip capacitor was placed across the ends of the coil and caused the coil to resonate at 247 MHz. The coil was

Figure 6.10. View of the polyvinyl chloride support surrounded by the silver-plated brass shield and located within saddle coil style modulation coils.

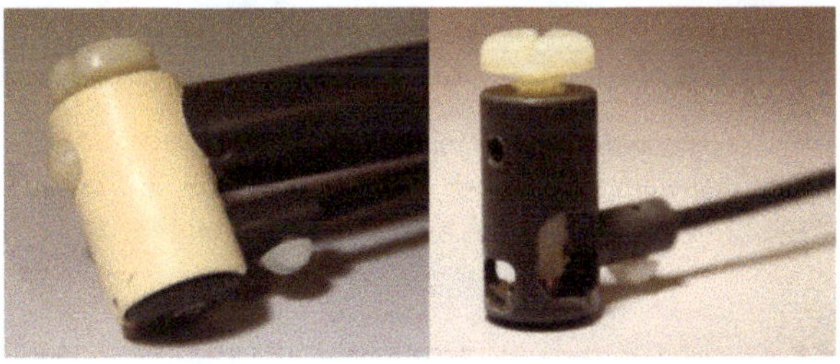

Figure 6.11. External views of the 2.9 mm surface coil. The left image is with the support arm and the right image is with it removed. Nylon screw controlled the coupling by varying the coupling loop to coil distance.

inductively coupled to the bridge via two turns of #26 copper wire placed axially to the coil, and a length of 50 Ω cable. The impedance was varied by changing the distance between the coupling loop and the *LC* coil via a nylon screw. See figure 6.11.

The 2 cm diameter coil was a seven-turn, microstrip, Archimedean spiral etched on a single-sided, 0.1143 mm thick, PTFE circuit board with 0.0381 mm thick copper. The equation in polar coordinates (r, θ) of the spiral is

$$r = a + (1/2)b + c\,\theta \tag{6.8}$$

where the radius of the inner spiral a = 4.65 mm, the width of the microstrip b = 0.5 mm, c = 0.115 mm, and the length $0 \leqslant \theta \leqslant 14\pi$ radians. The spiral had three gaps at 100, 500, and 645π radians along its length. The gaps at 500π and 645π were spanned by 1 pF chip capacitors. The resultant resonant frequency was 271 MHz. The spiral coil was inductively coupled to the 50 Ω cable of the spectrometer via one-turn of #18 copper wire in the form of a 1.5 cm diameter loop located coaxially above and opposite the sample side of the spiral. The impedance was adjusted using

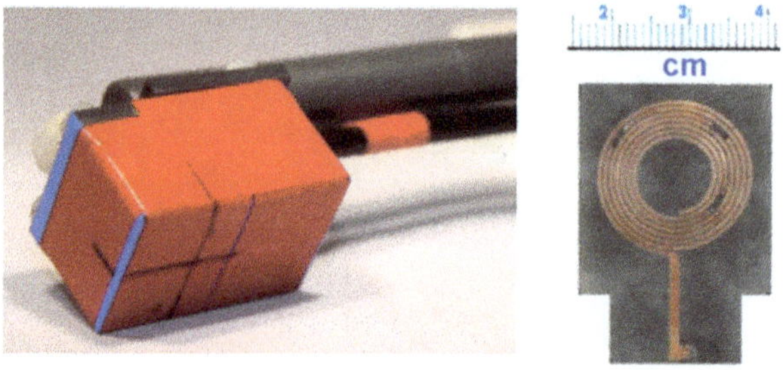

Figure 6.12. Images of the 2 cm spiral surface coil. The left image is an external view showing the bottom of the coil and the support arm. The right image presents the Archimedean spiral inductor.

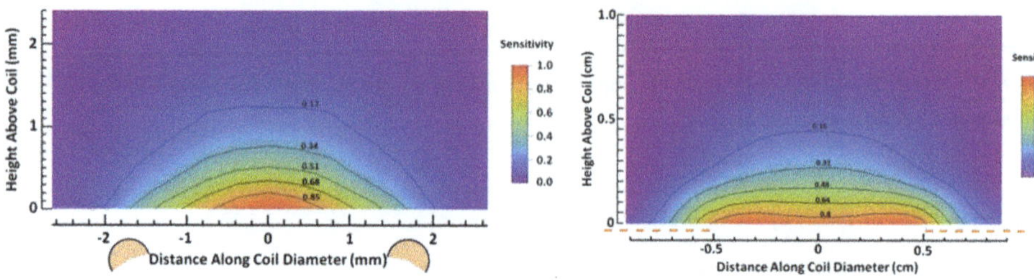

Figure 6.13. Relative spatial sensitivity of the 2.9 mm (left) and 2 cm (right) surface coils. The tan circles and dashed lines below the images represent the locations of the coil inductors. A wide-line low frequency electron paramagnetic resonance spectrometer. Reproduced from [50] with permission. © 2017 Wiley Periodicals, Inc.

a nylon screw (see figure 6.12) that slightly changed the angle between the coupling loop and spiral.

The relative spatial sensitivity of the two surface coils are depicted in figure 6.13. The sensitivity of the 2.9 mm surface coil drops to 0.5 by 0.5 mm from the surface. This equates to detecting signal from spherical caps adjacent to the surface coil. The 2 cm surface coil has a greater depth sensitivity extending the 0.5 sensitivity to approximately 1.5 mm adjacent to the coil. The sensitive area more approximates a 1 cm diameter disc.

The total signal from a surface coil is best understood by examining the sensitivity image for the 2.9 mm coil. The sensitivity drops off exponentially with distance from the center of the coil along y. (See figure 6.14.) The contours in xy are essentially circular meaning the contours of equal sensitivity can be defined in 3D as concentric spherical caps. The signal from spins in the contour interval 1.00 to 0.90 is much greater than that from spins in the 0.50 to 0.49 range, however the volume of the 0.50 to 0.49 range is much larger than that for the 1.00 to 0.99 range. For a finite volume of spatially uniform concentration spins adjacent to the coil, the lower sensitivities contribute significantly to the overall signal. The signal for a uniform concentration of spins in various thickness slabs of spins is presented in figure 6.14. For this 2.9 mm diameter surface coil the signal has reached 90% of its maximum with a 0.9 mm thick sample.

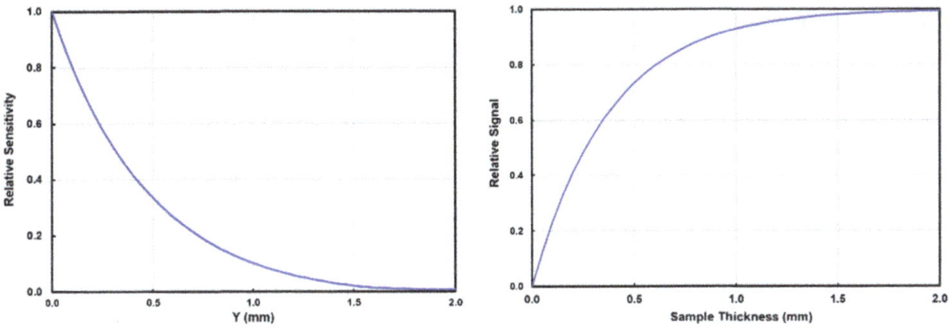

Figure 6.14. Plots of the relative sensitivity as a function of Y and relative signal as a function of uniform signal thickness for the 2.9 mm surface coil.

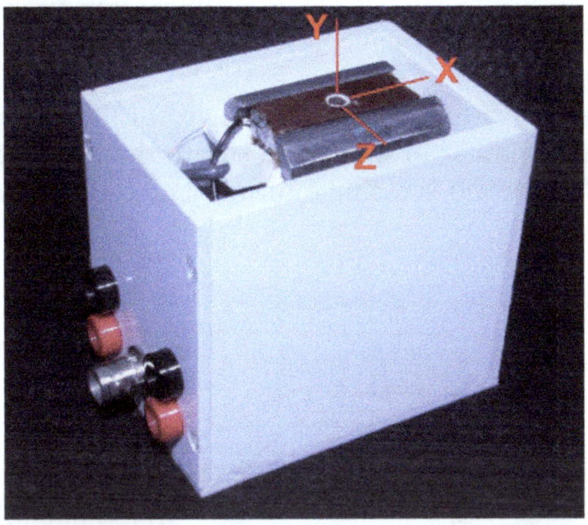

Figure 6.15. View of the top of the EPR MOUSE showing the unilateral magnet pole pieces and circular surface coil in the center. Reprinted from [49], Copyright (2017) with permission from Elsevier.

6.3.5 MOUSE

The EPR MOUSE replaces the magnet, modulation coils, and RF sample probe of the LFEPR spectrometer with a small unilateral magnet, modulation coils, and RF surface coil probe. The three weigh 560 g and are housed in a 59 × 74 × 82 mm polyethylene box shown in figure 6.15. The MOUSE utilizes all other spectrometer components except for the magnet power supply which is a 0–15 V, 0–6 A DC power supply (Kepco, TE 154-6M).

The geometry of the MOUSE magnet, probe, and their magnetic fields are described using a 3D Cartesian coordinate system where the origin of x, y, and z are at the surface of the magnet, halfway between and halfway along the length of the poles. The x-axis extends along the length and halfway between the poles. The z-axis points between the poles in the direction of the magnetic field and the y-axis extends upwardly from the top of the poles. The picture of the MOUSE in figure 6.15 shows the *xyz* coordinate system relative to the RF surface coil. Samples are centered at the origin of the *xyz* coordinate system and occupy the space where $y \geqslant 0$.

The yoke is constructed of 65 pieces of M15, 0.61 mm thick, cold-rolled non-grain-oriented (CRNGO) sheet steel laminated together to form a 4 cm long yoke. The magnetic field in the yoke and between the poles is induced by current sent through 420 turns of #26 enameled copper wire wound around each arm of the yoke. These are the sweep coils. A smaller set of two coils consisting of 76 turns of #32 enameled copper wire each were wrapped around the top of the two arms of the yoke to create the 10 kHz modulation field in the sample.

The yoke design was the result of several design iterations modeled in two dimensions (2D) using the Finite Element Method Magnetics (FEMM) package by David Meeker [67]. The 4 cm length of the yoke in x made the 2D model a reasonable approximation of our 3D device. The optimal single yoke geometry of the designs examined was one with sharp, flat-topped pole pieces as B_o decreased at a slower rate near the surface as y increased. Figure 6.16 presents the shape of the yoke in *yz*-plane.

We have experimented with two general designs for the RF surface coil. The first is a multi-turn solenoid similar in shape and size to the one described for use in the benchtop magnet. The second is a single turn solenoid design (STS).

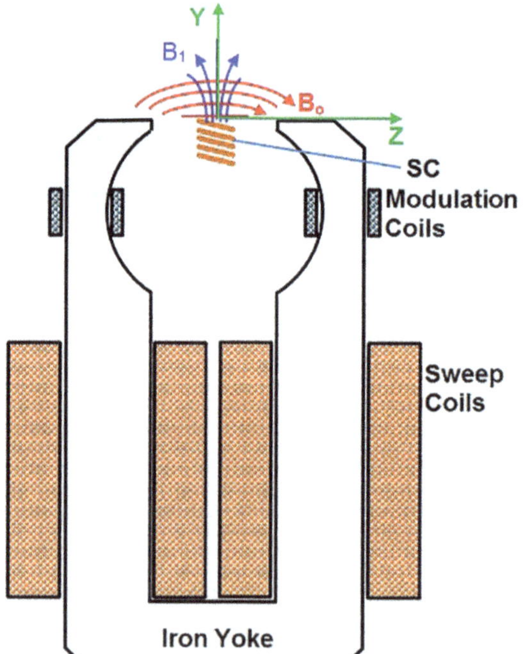

Figure 6.16. YZ cross sectional view through the EPR MOUSE showing the yoke, sweep coils, modulation coils, RF surface coil (SC), and the direction of the B_o and RF B_1 magnetic fields.

The multi-turn solenoid RF surface coil probe was a 2.9 mm inside diameter (ID), four-turn solenoid constructed of #26 enameled copper wire with a 5 pF capacitor across the inductor. The nominal resonance frequency and quality factor (Q) of the probe were 355 MHz and 71 respectively. The probe was inductively coupled to the 50 Ω cable of the spectrometer via a two-turn loop of #26 enameled copper wire. The coupling was varied by moving the coupling loop laterally over the solenoid in the x direction rather than changing the axial y distance. The coil was housed in a 22 × 12.7 × 35.5 mm HDPE block which was covered with 75 μm thick copper foil to form an RF shield around the coupling assembly and coil. A 5 mm diameter hole in the copper foil was centered on the multi-turn solenoid. This foil covered box fits between the pole pieces of the magnet such that the cylindrical axis of the multi-turn solenoid is coincident with the $-y$ axis and the end of the solenoid is in the xy plane.

The STS coil is constructed from 0.127 mm thick silver foil to form a 3.01 mm ID, 3.81 mm length silver foil cylinder with a 0.5 mm longitudinal gap running the length of the cylinder. The gap is spanned by five non-magnetic 22 pF chip capacitors (Knowles Capacitors, Cazenovia, NY). The STS is supported on the inside by a 3 mm OD, 7.5 mm length polyethylene rod and the outside by a 6.35 mm OD 3.26 mm ID Teflon cylinder. The STS, inside support, and outside support are secured together by cyanoacrylate glue. The STS is inductively coupled to the 50 Ω coaxial transmission line via a two-turn loop of #26 enameled copper wire wound around the polyethylene rod. The coupling is varied by altering the distance between the coupling loop and the STS. This relative position is fixed for most samples studied. The STS is surrounded by a silver-plated brass box measuring 8.89 × 8.89 × 25.4 mm, with a 6.35 mm hole in the 8.89 × 25.4 mm bottom for the Teflon STS support and a hole in an 8.89 × 8.89 mm side for the coaxial transmission line. The box is covered by an 8.89 × 25.4 mm silver-plated brass plate secured with four #0–80 brass screws. The nominal resonance frequency and quality factor (Q) of the STS probe were 420 MHz and 62 respectively. The brass box fits within a polyethylene support which in turn fits between the pole pieces of the magnet such that the cylindrical axis of the STS is coincident with the $-y$ axis and the end of the solenoid is in the xz plane.

The shield is an important part of the probe. The shield supports the resonator, increases the Q of the resonator, decreases the noise from external sources, and minimizes the detection of the undesirable low field RF absorption signal from the iron yoke. The latter signal can overwhelm any paramagnetic resonance signal from the sample.

Spectra are recorded by either placing the MOUSE on the sample, or the sample on the MOUSE. Since the yoke retains some magnetization from the previous scan, the yoke is demagnetized between scans by applying a decreasing 60 Hz sine wave to the main magnet coils.

6.3.6 Spectrometer control and data acquisition

National Instrument's LabVIEW code controlled the LFEPR spectrometer through a GPIB (IEEE-488) instrument bus. See figure 6.17 for a general diagram of the

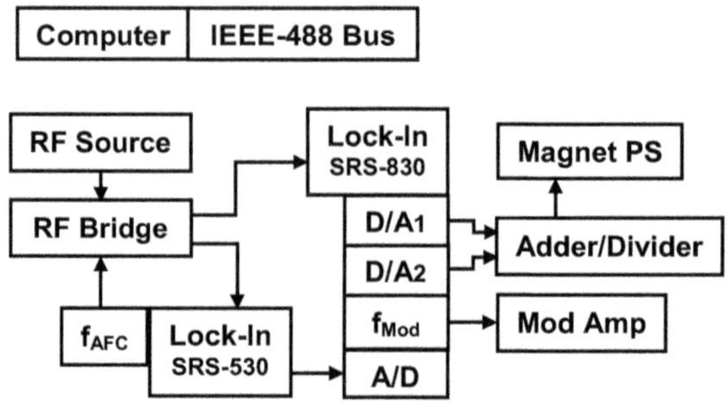

Figure 6.17. Block diagram of the control and data acquisition connectivity of the LFEPR spectrometer.

interconnections between the controlled devices of the LFEPR spectrometer. The RF source frequency is set either locally via buttons on the front panel or remotely by the program through the IEEE-488 bus. The main signal lock-in amplifier (SRS-830) can also be set locally or remotely over the bus, however data acquisition is only accomplished remotely. The LabVIEW code also sets voltages on two of the lock-in's digital-to-analog (D/A) converters. D/A_1 sets the sweep start and D/A_2 is used to step B_o through its sweep from $B_{o\mathrm{Min}}$ to $B_{o\mathrm{Max}}$ with a settable step size (δB_o) These two voltages are sent to a voltage adder and divider to convert the 0–10 V D/A signals to a zero to one-volt control voltage for the magnet power supply. This allowed the magnet power supply to be controlled to 0.1 mV in 10 V for a maximum of 10^4 steps in a B_o scan. The dwell time per acquired point (t_{DW}) is determined by some internal-to-the-program execution times and an externally supplied wait for magnet variable. The number of points in a spectrum is determined by $(B_{o\mathrm{Max}} - B_{o\mathrm{Min}})/\delta B_o$ and the total sweep time by $t_{\mathrm{DW}}(B_{o\mathrm{Max}} - B_{o\mathrm{Min}})/\delta B_o$ seconds.

This same signal lock-in also sets the B_o modulation frequency and amplitude. The LFEPR analog signal from the RF Bridge is detected by the SRS-830 lock-in and sent to the program via the IEEE-488 bus.

A second lock-in amp provides the 100 kHz AFC frequency to the RF bridge and detects the AFC error voltage from the bridge (V_E). This analog output signal is digitized by an analog-to-digital (A/D) input on the signal SRS-830 lock-in and sent to the LabVIEW program via the IEEE-488 bus. The AFC feature allows the user to control the AFC voltage threshold (V_T) above which the frequency is considered to have drifter and a change is required, and the frequency correction size and step direction ($\delta\nu$).

LabVIEW code given V_E performs the necessary logic to determine how to adjust ν, and sends the new ν value to the RF synthesizer before each acquired spectral data point. Before each LFEPR spectral point is recorded, a comparison is made between V_E and a user defined parameter threshold voltage (V_T). If $|V_E| > |V_T|$ action is taken to modify ν, if not, no action is taken and ν remains the same. The action taken to modify ν falls under two categories. If $V_E > V_T$, then ν is decremented by $\delta\nu$, and if $V_E < -V_T$, then ν is incremented by $\delta\nu$, the user defined correction frequency step size.

The LabVIEW pseudo code with symbols used in the chapter is

$$\nu = (i < 2)\ \nu: ((V_E > = V_T)\&\&(i > 1))\nu - \delta\nu: ((V_E < = -V_T)\&\&(i > 1))\nu + \delta\nu: \nu; \quad (6.9)$$

where i is the data point number in the spectrum. The operation of equation (6.9) starts with the second data point because before any spectral points are collected, the AFC code gives the sample probe a chance to settle on the optimum ν by iterating for 50 steps. Typical values for a 406 MHz, $Q = 440$ single turn solenoid probe are $V_T = 0.1$ V and $\delta\nu = 0.001$ MHz. Because the phase can vary from positive to negative or negative to positive as a function of ν upon going through resonance, both positive and negative inputs are acceptable for V_T and $\delta\nu$.

LabVIEW programs control the LFEPR spectrometer and allow it to perform four different acquisition functions. The programs give the operator control over different acquisition parameters. All programs allow the operator to set the number of points acquired and t_{Wait} at each point.

SPECTRUM performs a standard upfield magnetic field scan between $B_{o\mathrm{Min}}$ and $B_{o\mathrm{Max}}$ while recording the real output of the signal lock-in. The AFC is implemented in this program.

FIELD CYCLE performs cycled LFEPR where B_o is scanned from $B_{o\mathrm{Min}}$ to $B_{o\mathrm{Max}}$ and back to $B_{o\mathrm{Min}}$ a user defined set of times while recording the LFEPR signal from the signal lock-in amplifier. The AFC is implemented in this program. This type of acquisition is useful for examining ferro/ferrimagnetic samples.

TIME DOMAIN holds the B_o constant at a user defined value while recording the LFEPR signal. This program is used to determine the change in signal with respect to time. This can be used in kinetic studies if only the time is changing, and the spatial signal variation if the position of the sample probe relative to the sample changes with time.

LOCK-IN PHASE determines the optimum phase of the signal lock-in amplifier by recording multiple spectra differing in phase by a fixed number of degrees.

Table 6.3 summarizes input variables and their possible ranges in the four programs. A header in the top of the input file contains all the acquisition parameters set by the user.

6.4 Applications

6.4.1 Paintings

Paintings comprise a significant fraction of the objects humanity considers to possess cultural heritage significance. Even though most paintings are too large to be analyzed non-destructively in a conventional high frequency 9 GHz EPR spectrometer, the ability to see a signal from paramagnetic pigments in paint on canvas needed to be demonstrated. The next step was to demonstrate that a signal from pigments could be seen with LFEPR. Its higher limit of detection, small sweep width, and low field differences in the energy levels might prevent the detection of a signal. Assuming success with the latter, the following step would be to record a spectrum from paramagnetic pigments in paint on canvas with a surface coil in the 30 cm solenoidal magnet. The final step would be to utilize the EPR MOUSE, thus

Table 6.3. Range of program parameter values.

		Program			
Parameter	Symbol	Spectrum	Field cycle	Time domain	Lock-in phase
B_o sweep start	BoMin	0%–99%	Y	Y	0%–99%
B_o sweep stop	BoMax	1%–100%	Y	NA	1%–100%
B_o resolution	δB_o	10%–0.0001%	10%–0.0001%	NA	10%–0.0001%
Wait for magnet	t_{Wait}	0–100 s	0–100 s	NA	0–100 s
AFC threshold	V_E	y	y	y	n
AFC step	$\delta\nu$	0.001 MHz	y	y	n
Start phase	ϕ_{Min}	NA	NA	NA	0–179°
Stop phase	ϕ_{Max}	NA	NA	NA	1–180
Phase step	$\Delta\phi$	NA	NA	NA	0.1–180°
Output		Header B_o, Signal B_o, Signal B_o, Signal…	Header B_o, Signal B_o, Signal B_o, Signal…Optimum phase	Signal Signal Signal…	Header B_o, Signal B_o, Signal B_o, Signal…

demonstrating the ability to analyze pigments in paintings without a size limitation, thus achieving non-destructive analysis via EPR.

In the mid-1990s, Troup proposed the idea of using EPR to distinguish between paramagnetic pigments used in Renaissance paintings [68]. His approach was destructive in that it used a 9.1 GHz EPR and sample cooling to the temperature of liquid nitrogen, but it demonstrated the possibility on 14 different pigments. He proposed a modified microwave cavity would be possible that could examine slightly larger samples (~2 cm in length) non-destructively. His proposed design was never realized, but he did demonstrate the first step, observing an EPR signal from paramagnetic paint.

Subsequent steps were undertaken by the RIT Magnetic Resonance Laboratory. First, a review of paramagnetic pigments was undertaken to determine which signals had a measurable EPR spectrum at low frequencies. The second was demonstration of the LFEPR to detect a signal in dried oil-based paramagnetic pigment paint on canvas. The 30 cm diameter magnet and 2 cm surface coil would be used due to their better homogeneity and signal-to-noise ratio. The last was a demonstration that a useful LFEPR signal could be acquired using the EPR MOUSE. The results of these demonstration studies follow.

6.4.1.1 LFEPR of paramagnetic pigments

Many pigments are paramagnetic and possess an EPR signal. Many paramagnetic pigments do not possess an observable signal at room temperature because of their large linewidths. Cooling these pigments towards 4 K lengthens their relaxation time, decreases their linewidth, and allows detection of their EPR signal. The situation is more complicated with LFEPR because the sweep width is often less than the linewidth. Furthermore, sample cooling is not practical or possible due to the precious nature of some paintings. Tables 6.4 and 6.5 summarize those pigments for which a room temperature LFEPR signal has been observed. Included in the table are the literature *g*-factors and the LFEPR linewidths. Literature *g*-factors are reported instead of those measured by LFEPR due to the difficulty measuring *g*-factors from a fraction of a broad line. The Γ_{PP} of pigments with an LFEPR Γ_{PP} value less than the sweep width are reported. Γ_{PP} for pigments with a Γ_{PP} value greater than the sweep width are reported as >35 mT. Also presented in the table are other names for a pigment. Table 6.5 summarizes those pigments that are ferro/ferrimagnetic and identifiable by the hysteresis behavior of the LFEPR during a field cycled spectrum. All spectra were recoded of the pigments in bulk powder form using the STS resonator and the large solenoidal magnet. Acquisition parameters were optimized for the specific sample, i.e. modulation amplitude was smaller for smaller Γ_{PP} samples to avoid modulation broadening of a spectral line.

The pigment rhodochrosite is manganese carbonate ($MnCO_3$) and has an LFEPR spectrum from Mn(II). The LFEPR spectrum does not contain the over 22 resolvable LFEPR peaks in high frequency EPR or the large number visible in lower frequency EPR [60]. Instead it displays one large, homogeneously broadened Γ_{PP} = 32.1 mT Gaussian peak due to dipolar coupling between the manganese ions [69]. The

Table 6.4. Partial list of paramagnetic pigments with an LFEPR signal.

Name, chemical formula, and other names	Signal	*g* [ref]	LFEPR Γ_{PP} (mT)
Charcoal black	C˙	2.003 [77]	0.51
Coal black	C˙	2.0027 [78]	0.52
Carbon black Pflanzen schwarz (Gr), noir de charbon (Fr), nero carbone (It)	C˙	2.002 [73]	See text
Ultramarine $Na_{8-10}Al_6Si_6O_{24}S_{2-4}$ Lapis lazuli, Lapis, Cuprorivaite (Gr), Bleu outremer (Fr), blu oltremare (It)	$S_3^{-\cdot}$	2.027 74 [35]	2.24
Rhodochrosite $MnCO_3$ Manganese carbonate	Mn(II)	2.034 [69]	32.1
Marble dust	Mn(II)	$g_\perp$ = 1.9975–2.0018, $g_{//}$ = 2.0015–2.004, $g_{//}$ = 2.001 23, $g_{//}$ = 2.001 31 [60]	Many
Manganese sulfate tetrahydrate $MnSO_4{\cdot}4H_2O$	Mn(II)	1.9 [79]	17.7
Egyptian blue $CaOCuO(SiO_2)_4$ Ägyptisch Blau (Gr), bleu Egyptien (Fr), blue egiziano (It), Pompeian blue, Cuprorivaite, Calcium copper silicate	Cu(II)	$g_\perp$ = 2.054 $g_{//}$ = 2.326 [71]	10.86
Han blue $BaOCuO(SiO_2)_4$ Chinese blue	Cu(II)	$g_\perp$ = 2.093 $g_{//}$ = 2.383 [71]	2.23
Copper phthalocyanine $Cu(C_8H_4N_2)_4$ Phthalo blue	Cu(II)	2.052	13.6
Blue vitriol $CuSO_4{\cdot}5H_2O$ Copper sulfate, bluestone, vitriol of copper, Roman vitriol	Cu(II)	$g_\perp$ = 2.080 $g_{//}$ = 2.270 [80]	7.16
Copper sulfate monohydrate $CuSO_4{\cdot}H_2O$	Cu(II)	2.191	5.0
Copper sulfate $CuSO_4$	Cu(II)	2.24	6.6
$(NH_4)_2Cu(SO_4)_2{\cdot}6H_2O$ Copper ammonium sulfate hexahydrate	Cu(II)	g_z = 2·360 g_x = 2·071 g_y = 2·218 [81]	>25
Dioptase $CuSiO_3{\cdot}H_2O$ Copper silicate, chrysocolla	Cu(II)	$g_\perp$ = 2.0511 $g_{//}$ = 2.3601 [82, 83]	>35
Malachite $(Cu_2CO_3)(OH)_2$ Malachit (Gr), Bremen green, Olympian green, Iris green	Cu(II)	$g_{//}$ ~ 4, $g_\perp$ ~ 2 [84]	>35
Azurite $Cu_3(CO_3)_2(OH)_2$ Azurit (Gr), azurite (Fr), azzurrite (It)	Cu(II)	$g_{//}$ = 2.247 $g_\perp$ = 2.097 (Γ_{PP} ~ 7.6–10.2 mT) [85]	~15
Verdigris $Cu(CH_3COO)_2{\cdot}[Cu(OH)_2]_3{\cdot}2H_2O$ Basic Copper Acetate, Grünspan (Gr), Vertdegris (Fr), Verderame (It)	Cu(II)	$g_{//}$ = 2.29 $g_\perp$ = 2.056 [86]	>35
	Co(II)	2.22–2.71 (Γ_{PP} = 163–207 mT @ 9 GHz) [87]	»35

Cobalt green CoO-ZnO Rinmansgrün (Gr), vert de cobalt (Fr), verde cobalto, verde di zinco (It)			
Cobalt blue $CoAl_2O_4$ Cobalt(II) aluminateKobaltblau (Gr), bleu de cobalt (Fr), blu cobalto (It)	Co(II)	2.26 [76]	~17
Smalt $CoCO_3$, Cobalt glass	Co(II)		>20 mT @ 9 GHz
Potassium dichromate $K_2Cr_2O_7$	Cr(VI)		»35
Chromium ACAC Cr $(C_5H_7O_2)_3$ Chromium(III) acetylacetonate	Cr(III)	1.983 [88]	»35
Viridian green $Cr_2O_3{\cdot}2H_2O$ Chromoxidhydratgrün (Gr), vert émeraude (Fr), verde ossido di cromo (It)	Cr(III)	1.982 [89] 1.967 54 [90]	»35~50 @ 9 GHz
Chromium yellow $PbCrO_4$ Crocite (Complicated because of photodegradation to Cr(III) + Cr(V)) Chromgelb (Gr), jaune de chrome (Fr), giallo de cromo (It)	Cr(VI)	Cr(V) 1.980 Cr(III) 2.0, 4.0 [91, 92]	»35
Victoria green $Ca_3Cr_2(SiO_4)_3$ vert Victoria (Fr), SpaVictoria verde (Sp), Victoria grün (Gr), verde Victoria (It), Uvarovite, Vernalis, Victoria green garnet	Cr(III) Fe(III)	1.938 [72] (Impurity)	0.62 6.578 ± 0.027 [72]
1.05			
Terracotta red	Fe(III)	2.051 [93]	15.3
Anhydrous iron(II) sulfate $FeSO_4$	Fe(II)		7.5
$Fe(II)(NH_4)_2(SO_4)_2{\cdot}6H_2O$	Fe(II)	4.13	1.3
Sienna (Fe_2O_3, MnO)			
Silica glass beads	Fe(III)	4.3	0.9

homogeneous broadening causes the observed g-factor and Γ_{PP} in LFEPR spectroscopy to be dependent on ν.

Manganese sulfate tetrahydrate ($MnSO_4{\cdot}4H_2O$) is a Mn(II) transition metal pigment. It is similar to rhodochrosite in that only one homogeneously broadened peak is observed, however the linewidth is narrower at $\Gamma_{PP} = 17.7$ mT.

Ultramarine blue ($Na_8Al_6Si_6O_{24}S_3^{\cdot}$) is a synthetic version of the mineral Lapis lazuli. Its EPR signal is associated with a $S_3^{-\cdot}$ radical. The LFEPR g-factor is approximately equal to the literature value of 2.027 74 [35]. Γ_{PP} is frequency dependent, and the LFEPR value of 2.24 mT is within the literature range [70].

Table 6.5. Ferro/ferrimagnetic pigments partially identified by LFEPR hysteresis.

Name, chemical formula, and other names	Signal
Prussian blue	Fe(II)Fe(III)
$Fe_4[Fe(CN)_6]3 \cdot xH_2O$ x = 14–16	
Iron(III)–hexacyanoferrate(II)	
Berliner Blau (Gr), bleu prussien (Fr), blue di Prussia, azzuro di Berlino (It), Berlin blue, Parisian Blue, Paris Blue, Turnbull's blue, Milori blue, iron blue	
Magnetite	Fe(II)Fe(III)
Fe_3O_4	
Red Ochre	Fe(III)
Fe_2O_3, Hematite	
Roter Ocker (Gr), Ocre rouge naturelle (Fr), Ocra rossa (It)	
Brown Ochre	Fe(III)
FeO(OH), $Fe_2O_3 \cdot H_2O$, Goethite	
Yellow Ochre	Fe(III)
$FeO(OH) \cdot nH_2O$, Iron oxyhydroxide, Gelber Ocker (Gr), ocre jaune (Fr), ocra gialla (It), mars yellow, limonite	
Burnt Sienna (iron oxide and manganese oxide)	Fe(III), Mn(II)

Egyptian blue ($CaOCuO(SiO_2)_4$) and Han blue ($BaOCuO(SiO_2)_4$) are isomorphs, differing only in the presence of a non-paramagnetic calcium or barium ion. They have similar literature $g_{//}$ and $g_{\perp}$ values [71], although little asymmetry is seen in LFEPR spectra, but different linewidths with Egyptian blue being approximately five time broader than Han blue.

Victoria green ($Ca_3Cr_2(SiO_4)_3$) displays two peaks but they are believed to be from impurities in the calcium chromium silicate. The $g \approx 3.567$ is from Fe(III) and the $g \approx 1.938$ is from Cr(III) [72]. If these peaks are present in all Victoria green they can be used as a marker for this pigment regardless of their origin.

Copper phthalocyanine ($Cu(C_8H_4N_2)_4$), also known as phthalo blue, is a Cu(II) centered phthalocyanine complex. It is a modern era pigment (1927) and is considered an anthropogenic organic pigment even though it is copper based. It has a $g = 2.052$ and a 13.6 mT Γ_{PP}.

Terracotta red is a red pigment derived from fired clay with a high iron content. The sample is predominantly paramagnetic with a small amount of ferro/ferrimagnetism as evidenced by field cycling LFEPR. The color is most likely from Fe_2O_3, but the predominant EPR signal is not as Fe_2O_3 is not paramagnetic. Perhaps for this reason not all terracotta red pigments possess the same EPR signal. Our sample was derived from a crushed terracotta red flower pot and had a broad, asymmetric, and intense 15.3 mT linewidth characteristic of a powder pattern spectrum with a $g_{\perp} \approx 2.03$ and $g_{//} > 3.8$. Since terracotta red is not a pure substance there is no literature value for the g-factor.

Charcoal black and coal black are carbon free radical pigments derived from charcoal and coal, respectively. Charcoal is made from the pyrolysis of plant-based matter. The process forms some stable carbon-centered radicals with a $g = 2.003$ and $\Gamma_{PP} \approx 0.5$ mT. Coal also contains carbon-centered radicals from the anaerobic decomposition of plat matter under pressure and heat. The g-factor and Γ_{PP} are similar to those for charcoal. The concentration of free radicals can vary significantly between coal types. Carbon black is a pigment manufactured by the incomplete combustion of hydrocarbons. It may possess carbon-centered radicals and an EPR signal, but not all carbon black does [73]. Those that do tend to have oxygen sensitive linewidths.

Blue vitriol ($CuSO_4 \cdot 5H_2O$) is more commonly known today by its chemical name copper sulfate pentahydrate. The pentahydrate has an intense, slightly asymmetric, broad 7.16 mT peak on the LFEPR spectrometer. Oven drying of an oil-based paint of blue vitriol will change colors with drying temperature between room temperature and 100 °C. This is thought to be due to a reaction between the linseed oil and the Cu (II) [74, 75]. Heating this pentahydrate to 100 °C will cause the loss of two waters of hydration. Heating to 140 °C will cause the loss of another two water molecules and results in a more intense signal with a $\Gamma_{\mathrm{PP}} \approx 5$ mT. The last water of hydration is lost by 250 °C to form anhydrous $CuSO_4$ without an EPR signal.

Marble dust pigment like marble possesses an EPR signal from trace amounts of Mn(II) substituted for Ca(II) in the calcium carbonate lattice of calcite or calcium magnesium carbonate lattice of dolomite. The multiple peaks at low frequencies [60] and the signal being from impurity concentrations of Mn(III) make the LFEPR signal difficult to discern. Please refer to the final application in this chapter on marble for more information.

Cobalt blue with chemical name cobalt(II) aluminate ($CoAl_2O_4$) is reported to have a $g = 2.26$ and a broad linewidth of several Tesla at a few Kelvin [76]. The LFEPR spectrum of cobalt blue has a similar g-factor and $\Gamma_{\mathrm{PP}} \approx 17$ mT. These two observations are inconsistent and perhaps the pigment signal is an impurity.

Many pigments possess an LFEPR signal but the signal is too broad to effectively identify from a spectrum on the current LFEPR spectrometer. Some of these are copper ammonium sulfate hexahydrate, dioptase, malachite, verdigris, cobalt green, potassium dichromate, and chromium ACAC, veridian green, and chromium yellow. These $g \approx 2$ systems possess a Γ_{PP} greater than the current sweep width of the LFEPR spectrometer solenoid magnet. An example of this point is veridian green which is reported to have a $\Gamma_{\mathrm{PP}} = 50$ mT at 9 GHz. This large Γ_{PP} puts the low field inflection point of a first derivative signal hypothetically less than $B_o = 0$ zero and the high field inflection point beyond the sweep width of the spectrometer. The result is a spectrum with a negatively sloping baseline. These samples with a confirmed high frequency EPR signal are reported in table 6.4 as having a LFEPR $\Gamma_{\mathrm{PP}} > 35$ mT.

A group of several common pigments are either ferromagnetic or ferrimagnetic, and as such do not possess a paramagnetic resonance signal. Their presence can be identified by field cycling LFEPR spectroscopy as presented in figure 6.5 of the theory section. They all contains iron and are all related to some extent.

Hematite (Fe_2O_3) is a deep, dark red pigment made from grinding hematite rock. It is perhaps one of the earliest pigments; found in 40 000 year old cave paintings. It is not a pure material and as such can contain many other iron compounds. Fe_2O_3 can exist in at least four different structures identified as α-, β-, ε-, and γ-Fe_2O_3, but α and γ are the most common and stable. α-Fe_2O_3, has a rhombohedral lattice and it is slightly ferromagnetic. γ-Fe_2O_3 has a cubic lattice and is ferromagnetic. Our samples display ferromagnetic properties between 20 °C and 500 °C, and a very small $g \approx 4$ paramagnetic resonance signal between 600 °C and 1200 °C.

Ochres are a class of Fe(III) oxide pigments derived from natural earth deposits. As such they also contain clay and sand. Yellow Ochre ($FeO(OH){\cdot}nH_2O$) contains an iron oxyhydroxide of Fe(III). Heating yellow ochre drives of water first producing brown ochre($FeO(OH)$, $Fe_2O_3{\cdot}H_2O$). Further heating produces red ochre (Fe_2O_3). All ochres possess some ferromagnetism.

Magnetite (Fe_3O_4 also written as $Fe(II)Fe(III)_2O_4$) is a ferrimagnetic, black, iron oxide pigment containing Fe(III) and Fe(II). Prussian blue ($Fe(iii)_4[Fe(ii)(CN)_6]_3{\cdot}xH_2O$ x = 14–16), or iron(III)–hexacyanoferrate(II), is similar to magnetite in that it contains Fe(II) and Fe(III) and is ferrimagnetic.

6.4.1.2 Paint on canvas

Linseed oil-based paints of eight pigments on canvas were studied by Javier [93]. Ultramarine blue (UM), $Na_8Al_6Si_6O_{24}S_3^{\cdot}$, is a sulfur-centered $S_3^{-\cdot}$ radical containing pigment. Egyptian blue, $CaCuSi_4O_{10}$, and Han blue or effenbergerite, $BaCuSi_4O_{10}$, are similar copper (II) ion complexes differing only in the presence of calcium or barium. Blue vitriol, $CuSO_4{\cdot}5H_2O$, has a broad EPR signal from copper (II). Rhodochrosite, $MnCO_3$, possesses one very broad EPR absorption due to exchange broadening from the close proximity of the manganese (II) ions. Anthracite coal and charcoal are black pigments containing carbon-centered radicals. Terracotta red is a red pigment from fired iron (III) containing clay.

Each pigment was mixed with boiled linseed oil and a suspension created using either a small mechanical mixer for five minutes or a mortar and pestle. The target pigment volume concentration (PVC) for each paint was the critical pigment volume concentration (CPVC) based on observed physical properties of the suspension [94]. The CPVC was arrived at empirically and iteratively by adding either more linseed oil or pigment. The paint suspension was applied to a piece of cardboard backed titanium dioxide primed canvas to form a paint swatch. One thick and one thin swatch was prepared for each pigment. The paint swatches ranged in thickness from 40 to 860 μm. Table 6.6 lists the thickness of the paint swatches.

LFEPR spectra were recorded using the 2 cm diameter surface coil. Acquisition parameters were optimized for the specific sample, i.e. the modulation amplitude was smaller for smaller Γ_{PP} samples to avoid modulation broadening of a line. No LFEPR signal was associated with the canvas or primer. The literature *g*-factors and LFEPR linewidths of these pigments can be found in table 6.4. Figures 6.18–6.21 present the spectra from the eight paint pigments and table 6.6 summarizes the signal-to-noise ratio (SNR) values for the thin and thick swatches for each pigment paint. Use of the same B_o scale allows comparison of figures 6.18–6.20.

Table 6.6. Summary of the SNR values for different thickness paint swatches.

		Thin swatch		Thick swatch	
Pigment in dried linseed oil paint	Γ_{PP} (mT)	Thickness (μm)	SNR	Thickness (μm)	SNR
Han blue	2.23	170	104	400	294
Ultramarine blue	2.24	120	78	580	216
Blue vitriol	7.16	270	160	710	733
Egyptian blue	10.86	350	30	360	49
Charcoal black	0.51	40	<1	450	55
Coal black	0.52	120	<1	860	1.35
Terracotta red	15.3	250	2045	500	2110
Rhodochrosite	32.1	200	170	680	366

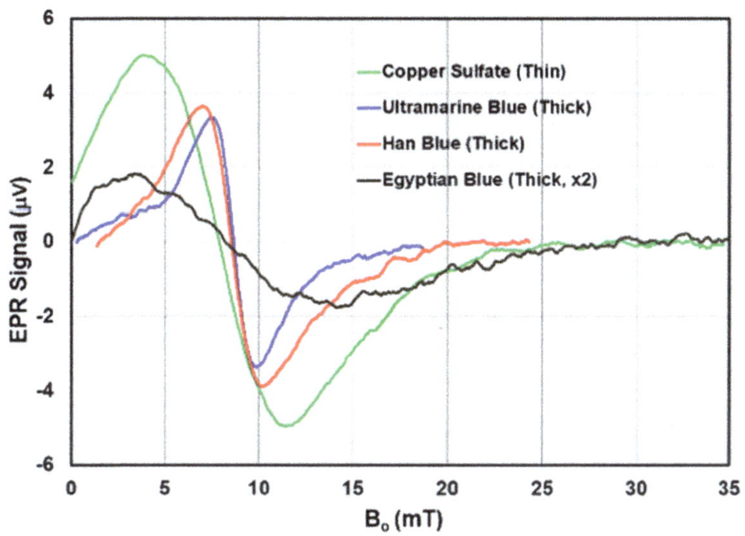

Figure 6.18. LFEPR spectra of the blue pigment paints on canvas recorded with the 2 cm diameter surface coil at 270 MHz. See text for an explanation of thick and thin samples.

In all cases, the sample was within the >70% sensitivity profile of the surface coil. The thicker the sample, the better the SNR, but the relationship between the two is not linear as the sensitivity profile of figure 6.12 drops off nonlinearly with distance from the coil. Since the 2 cm diameter surface coil is optimized for a 1 cm diameter disc shaped area, a smaller surface coil may be necessary for examining smaller paint regions.

The LFEPR spectrometer and the 2 cm surface coil operating at 270 MHz was capable of detecting a signal for each of the eight pigments tested. The most challenging pigments were charcoal and coal as the thinner swatches had an SNR less than one. Clearly, the thicker swatches provided a better signal.

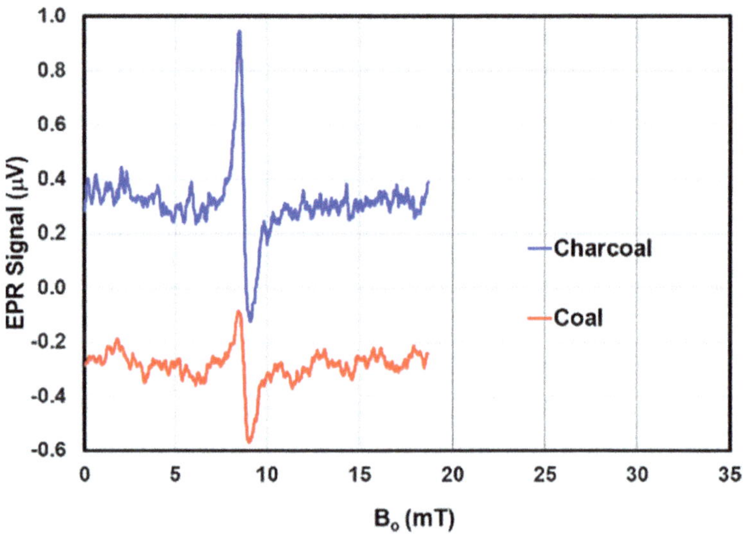

Figure 6.19. LFEPR spectra of the black pigment paints charcoal and coal on canvas recorded with the 2 cm diameter surface coil at 270 MHz.

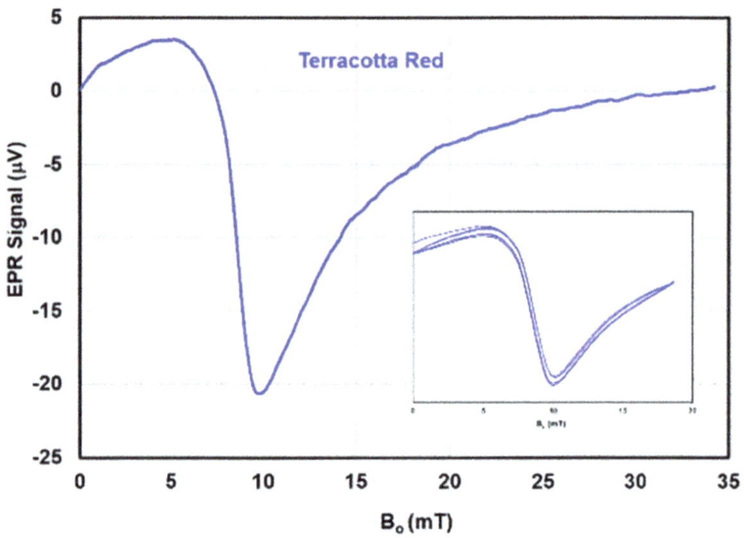

Figure 6.20. LFEPR spectrum of terracotta red pigment paint on canvas recorded with the 2 cm diameter surface coil at 270 MHz. Insert is a field cycle scan showing slight ferromagnetism in the sample.

The Γ_{PP} of rhodochrosite is comparable to the 35 mT sweep width and therefore difficult to recognize. Fitting the absorption spectrum of the first derivative line with a Gaussian line shape allowed the calculation of the linewidth and *g*-factor. The full width of the absorption and first derivative lines are shown to guide the eye, as scanning $B_o < 0$ would give the mirror image of the $B_o > 0$ spectrum. The 200 and

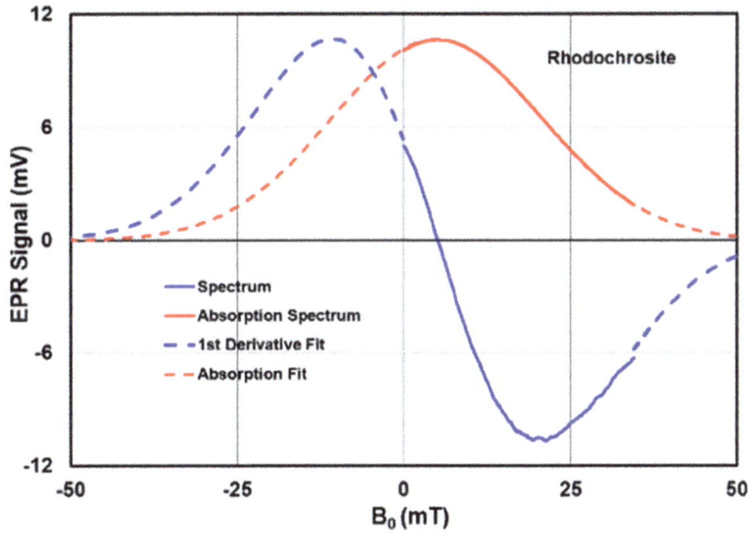

Figure 6.21. Absorption and first derivative LFEPR spectra of rhodochrosite at 270 MHz. Solid line is the actual spectrum and its integration. The dashed lines represent the line shape beyond the field sweep. See text for detail.

680 mm thick samples had a SNR of 170 and 360, thus allowing easy identification of rhodochrosite.

There were four blue pigments studied. Ultramarine blue has a respectable SNR of 78 and 216 for the 127 and 584 μm swatches respectively. Γ_{PP} is independent of sample thicknesses and its value of 2.24 mT is within the literature value range [70]. This pigment is easy to identify by LFEPR based on the *g*-factor, Γ_{PP}, and SNR. Blue vitriol possessed a very intense $\Gamma_{PP} = 7.16$ mT LFEPR absorption showing very little anisotropy. Han blue and Egyptian blue have similar *g*-factors but different linewidths. Han blue has a narrower 2.23 mT linewidth compared to Egyptian blue's 11.19 mT value. All four of these pigments are readily distinguished from each other.

The linewidths of coal and charcoal are the narrowest studied with the surface coil making the signals easy to identify. Their smaller peak areas indicate that the electron spin concentrations are very small in comparison to that of the other samples.

Terracotta red pigment in the linseed oil paint has a SNR > 2000 for both the thick and thin swatches, making it the largest signal of the pigments studied with the surface coil. The spectral line shape is very asymmetric, attributable to an anisotropic *g*-factor and a small amount of ferromagnetism. Terracotta red is not a pure substance making its properties variable from source to source.

The eight pigments studied are examples of classical pigments in linseed oil on canvas. Results will be similar for any pigment with a LFEPR signal in a 1 cm diameter region provided it will fit into the magnet. Smaller diameter paint regions can be analyzed with the 2.9 mm diameter surface coil [50], however the SNR will be lower. These eight pigments should yield similar results in larger paintings when the EPR MOUSE is utilized [49]. Other pigments with an LFEPR signal and a $\Gamma_{PP} \lesssim 32$ should

Table 6.7. LFEPR spectral acquisition parameters.

Parameter	Value
Spectral points	1000
Dwell time	215 ms pt^{-1}
Time constant	0.3 s
B_o sweep start	0.05 mT
B_o sweep end	50.48 mT
B_o step size	0.050 48 μT
Operating frequency (ν)	421.8 MHz
B_o modulation frequency	10 kHz
B_o modulation amplitude	0.22 mT
Number of averages	4–10

yield a detectable signal as a paint on canvas with this surface coil approach; however, there is no comprehensive database of pigments displaying an LFEPR signal. Every pigment identified in table 6.4 with $\Gamma_{PP} < 32$ mT and table 6.5 should be identifiable by LFEPR. The ability of LFEPR to detect an EPR spectrum for other pigments must be determined experimentally. All the swatches in this study were of single pigment paints. The next study in this entry demonstrated the ability of the EPR MOUSE to identify pairwise mixtures of seven pigments in linseed oil paint.

6.4.1.3 Paint mixtures on canvas

Many artists work with a small set of primary pigment colors from which all the colors in their pallet are made. The French painter Oscar-Claude Monet is a noted example of this [95]. Therefore it is not only necessary for an analytical technique to be able to identify the pigment in a single pigment paint, but also mixtures of pigments in a paint. Spectral separation of mixtures is routinely done in EPR [96]. Bogart *et al* [97] demonstrated the ability of the EPR MOUSE to identify 21 pairwise mixtures of charcoal, ultramarine, Han blue, Egyptian blue, blue vitriol, terracotta red, and rhodochrosite pigments in paint on canvas. Paints were prepared by combining approximately equal volumes of CPVC paints. These mixtures were not quantitative due to the difficulty of preparing and mulling small (~1 ml) quantities quantitatively. Mixed paints were applied to titanium dioxide primed canvas on a cardboard backing to form swatches. Each swatch was greater than 7 mm in diameter and varied in thickness. The thickness of each swatch was between 1.1 and 2.3 mm. No accelerators were used to facilitate auto-oxidation and crosslinking of the linseed oil in the paint swatches. The auto-oxidation was either allowed to occur by itself in air at room temperature over a week, or by heating in an oven at 60 °C for approximately 24 h. In addition to the 21 mixture swatches, swatches of the seven pure pigments in linseed oil were prepared.

Spectra of the swatches were recorded using the EPR MOUSE and the LFEPR spectrometer acquisition parameters listed in table 6.7.

Table 6.8. Spectral properties of the seven library spectra.

Pigment	Zero crossing (mT)	Γ_{PP} (mT)
Ultramarine blue	14.79	2.56
Han blue	14.53	6.66
Egyptian blue	13.96	10.85
Terracotta red	13.05	6.19
Rhodochrosite	10.48	23.63
Charcoal	15.07	0.92
Blue vitriol	13.75	7.31

Table 6.9. Calculated f_i values for single component swatches.[a]

	Components						
Swatch spectra	Charcoal	Terracotta red	Han blue	Egyptian blue	Rhodochrosite	Ultramarine	Blue vitriol
Charcoal	0.99	0.00	0.00	0.00	0.00	0.00	0.00
Terracotta red	0.00	0.99	0.00	0.00	0.00	0.00	0.01
Han blue	0.00	0.00	1.00	0.00	0.00	0.00	0.00
Egyptian blue	0.00	0.00	0.00	0.99	0.01	0.00	0.01
Rhodochrosite	0.00	0.00	0.00	0.04	0.96	0.00	0.00
Ultramarine	0.01	0.00	0.00	0.00	0.00	0.99	0.00
Blue vitriol	0.02	0.00	0.02	0.04	0.00	0.00	0.92

[a] f_i values rounded to the nearest 0.01.

The seven single pigment paints had LFEPR spectra with the zero crossing and Γ_{PP} values listed in table 6.8. These are presented so show the variety of narrow and broad spectral peaks, and *g*-factors for the pure paints. The LFEPR of broad spectra can have zero crossings which do not correspond to the *g*-factor. Therefore, similar zero crossings make it more difficult to resolve spectral components.

The seven single pigment spectra became the library spectra for a fitting algorithm. The analysis determined the optimum combination of the seven library spectra needed to achieve the best fit of a mixture spectrum. The algorithm minimized the square of the differences between the actual and calculated spectra by varying the amounts of the seven library spectra. The output of the algorithm is a fraction (f_i) of each of the seven library spectra in a mixture spectrum. A first test of the algorithm was how well it analyzed spectra from the single component swatches. Separate single component swatch spectra were recorded and fit with the seven library spectra. Table 6.9 summarizes these results by presenting for each single component swatch spectrum the fractions of the seven library spectra needed to produce a best fit. In all seven cases, the major component of each of the seven single pigment swatch spectra was the actual pigment with $f_i \geqslant 0.92$, and five with $f_i \geqslant 0.99$.

In the two cases where $f_i = 0.92$ (blue vitriol) and 0.96 (rhodochrosite), the need for other minor library components is attributed to low-frequency noise in a spectral acquisition which is fit by $f_i < 0.04$ of components not present in the mixture.

Results for spectra from the 21 pairwise mixture swatches are presented in table 6.10. Figure 6.22 presents the spectra for the blue-vitriol swatch. The red line represents the spectrum from the mixture swatch and the blue the best algorithm fit based on adding the amount of blue vitriol and charcoal spectra represented by the

Table 6.10. Calculated f_i values for two-component paint swatches.[a]

	Components						
Swatch spectra	Charcoal	Terracotta red	Han blue	Egyptian blue	Rhodochrosite	Ultramarine	Blue vitriol
Rhodochrosite + Charcoal	0.68	0.00	0.00	0.00	0.32	0.00	0.00
Han blue + Charcoal	0.66	0.00	0.19	0.05	0.00	0.07	0.03
Egyptian blue + Charcoal	0.32	0.00	0.00	0.68	0.00	0.00	0.00
Terracotta red + Charcoal	0.43	0.55	0.00	0.00	0.00	0.01	0.01
Rhodochrosite +Terracotta red	0.00	0.61	0.00	0.00	0.38	0.00	0.01
Han blue + Terracotta red	0.00	0.67	0.33	0.00	0.00	0.00	0.00
Egyptian blue + Blue vitriol	0.00	0.00	0.00	0.70	0.02	0.00	0.28
Ultramarine + Blue vitriol	0.00	0.00	0.00	0.27	0.05	0.03	0.65
Ultramarine + Han blue	0.00	0.00	0.60	0.00	0.00	0.40	0
Blue vitriol + Charcoal	0.43	0.00	0.00	0.00	0.00	0.00	0.57
Ultramarine + Charcoal	0.53	0.00	0.00	0.00	0.01	0.46	0.00
Rhodochrosite + Han blue	0.00	0.02	0.37	0.06	0.51	0.05	0.00
Egyptian blue + Rhodochrosite	0.00	0.01	0.00	0.40	0.58	0.00	0.01
Blue vitriol + Rhodochrosite	0.00	0.02	0.00	0.00	0.44	0.00	0.55
Rhodochrosite + Ultramarine	0.00	0.03	0.01	0.00	0.83	0.14	0.00

Han blue + Egyptian blue	0.00	0.00	0.49	0.27	0.05	0.00	0.19
Han blue +Blue vitriol	0.00	0.00	0.29	0.00	0.03	0.04	0.64
Egyptian blue + Terracotta red	0.03	0.63	0	0.31	0.00	0.00	0.03
Egyptian blue + Ultramarine	0.00	0.01	0.02	0.56	0.02	0.28	0.10
Terracotta red + Blue vitriol	0.00	0.59	0.00	0.00	0.00	0.00	0.41
Terracotta red + Ultramarine	0.00	0.78	0.00	0.00	0.00	0.16	0.06

[a] f_i values rounded to the nearest 0.01.

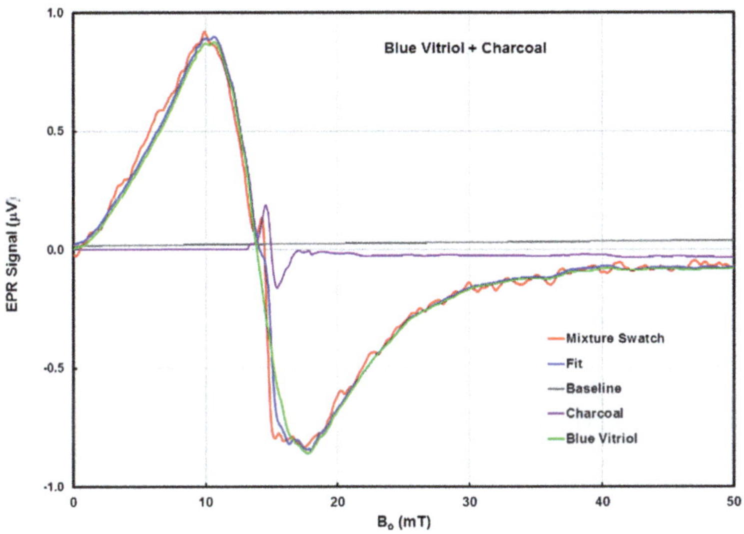

Figure 6.22. Spectra used in the fitting of the spectrum from the swatch containing a mixture of blue vitriol and charcoal. Spectra labled baseline, charcoal, and blue vitriol represent the components used to create the fit spectrum.

green and purple spectra respectively. The gray line is the amount of linear baseline drift needed to fit the mixture spectrum. The algorithm correctly identified the top two pigments in the swatch as charcoal and blue vitriol. The values of f_i for the components of the 21 mixtures are presented in table 6.10.

In all cases, the algorithm correctly identified the top two pigments in the paint swatches. In some cases, the algorithm required some minor additional components to achieve an optimal fit. This was attributed to low-frequency noise in the spectrum which was best fit by adding small amounts of other library spectra. Quantitative

Table 6.11. Calculated f_i values for layered paint swatch.[a]

	Components						
Swatch spectrum	Charcoal	Terracotta red	Han blue	Egyptian blue	Rhodochrosite	Ultramarine	Blue vitriol
Ultramarine over charcoal	0.37	0.03	0.00	0.00	0.00	0.60	0.00

[a] f_i values rounded to the nearest 0.01.

EPR is challenging [98], and quantitative EPR with the surface coil and unilateral magnet found in the EPR MOUSE is more so. For this reason, they did not focus on attaining $f_i \approx 0.5$ for two-component mixtures.

These results demonstrate the ability of the MOUSE and the algorithm to correctly identify the two pigments in a two-component paint. Because the EPR MOUSE detects signal from as deep as 1.5 mm, this approach is also suitable for identifying pigments in underlying paints not visible to the human eye. A demonstration of this is presented next.

6.4.1.4 Paint layer analysis

Underpaintings are commonly found underneath visible paintings. These are generally not detectable by optical spectroscopies that only examine the surface. LFEPR, like some other analytical techniques, are bulk analytical techniques capable of recording an EPR signal from an entire 1.5 mm thick sample.

In this study a 0.46 mm layer of ultramarine blue paint was placed over a 1.83 mm layer of charcoal black paint located on a canvas backing. The charcoal layer was allowed to harden before the ultramarine layer was added. This eliminated the possibility of diffusion of carbon black into the upper ultramarine layer, and ultramarine into the carbon layer.

Table 6.11 summarizes these results. Based on the sensitivity profile of the surface coil of the MOUSE, it is expected that the upper ultramarine layer would contribute the most to the signal and charcoal a lesser amount. This conjecture was confirmed. Like many of the other analyses, a lesser component was found which was not present in the swatch. In this case it was terracotta red and also attributed to noise in the spectrum.

6.4.2 Ceramics

Many clays contain transition metal ions that, depending on their oxidation state and local environment, can be paramagnetic. By far the most prevalent of these is iron, which can also be ferro/ferrimagnetic. The LFEPR signal can be used to not only identify the presence of iron and if it is ferro/ferromagnetic or paramagnetic, but also the firing temperature of the clay and the formulation of the clay.

Figure 6.23. Images of red art clay samples as a function of firing temperature between 100 °C and 1200 °C.

Ceramics are classified in part by their firing temperature (T_F). Earthenware has the lowest firing temperature of 600 °C–1200 °C, followed by stoneware between 1100 °C and 1300 °C, and porcelains between 1200 °C and 1400 °C.

6.4.2.1 Firing history

Red art clay contains 7% Fe_2O_3 by weight. When fired at different temperatures between 100 and 1200 °C it tales on different colors and physical properties. (See figure 6.23.) Since the red color of the clay is attributed to the iron, the chemical properties of the iron must be changing with increased firing temperature.

For insight into the chemistry, consider the following study of these clay samples. One was red art clay with 7% Fe_2O_3. A second was kaolin clay containing 0.7% Fe_2O_3. Three other samples were mixtures of kaolin and 3.5%, 7%, and 15% Fe_2O_3. Clay samples were hydrated with deionized water, mixed, kneaded, and rolled into 1 cm diameter by 65 cm long rods. The clay rods were dried at 100 °C for 2 h, and then cooled to room temperature before firing. Samples were fired at 12 different temperatures between 100 °C and 1200 °C. LFEPR spectra were recorded at 440 MHz using a STS sample probe. Figure 6.24 presents the LFEPR spectrum for the 300 °C sample of red art clay. This spectrum depicts two of the three characteristic features of fired red art clay spectra. The first is a narrow $g \approx 2$ peak possibly from f-centers, the second is a broad $g \approx 4.3$ peak from high spin Fe (III) in a rhombic environment [99], and the third is evidence of ferro/ferrimagnetism from field cycling (figure 6.25). Note the small $g \approx 4.3$ paramagnetic resonance peak in figure 6.25.

Figure 6.26 shows the dependence of the peak-to-peak first derivative signals (S_{PP}) as a function of firing temperature for red art clay. The $g \approx 2$ peak increased by ~80% between 100 °C and 270 °C, and then decreased, disappearing completely by 500 °C. The $g \approx 4$ peak increases in amplitude as firing temperature increases. At 850 °C the sample develops ferro/ferromagnetic properties which disappear above 1050 °C.

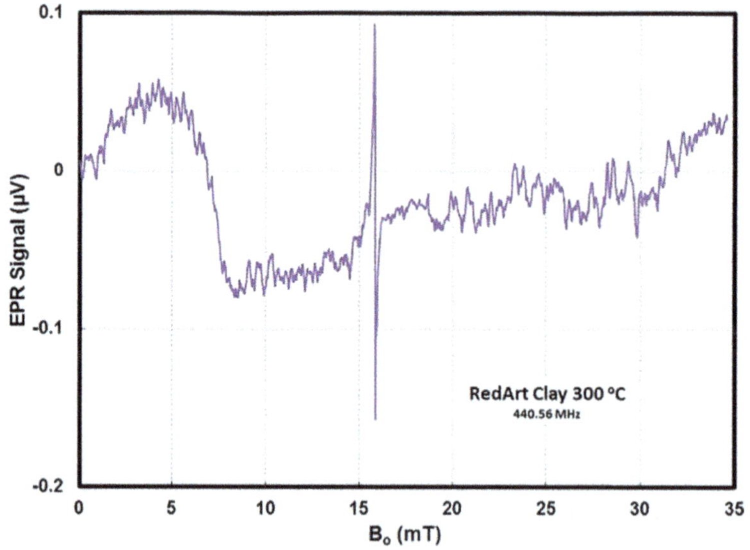

Figure 6.24. LFEPR spectrum of red art clay at 300 °C showing a narrow $g \approx 2$ peak possibly from f-centers, and a $g \approx 4.3$ peak from Fe(III).

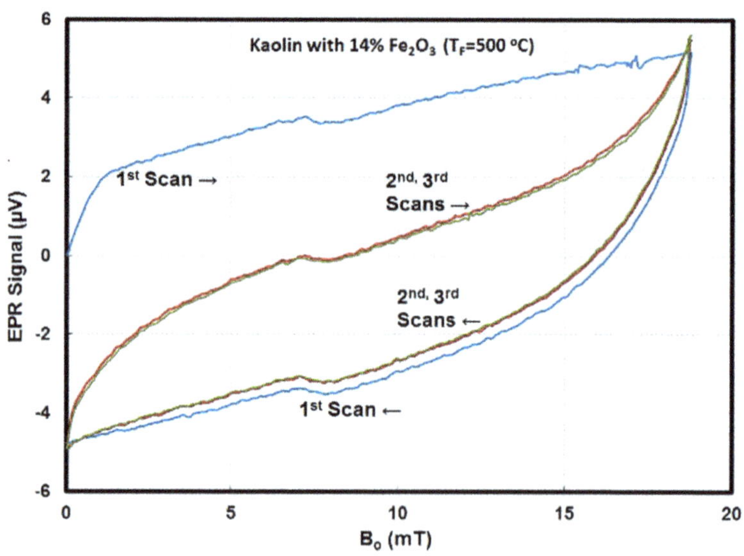

Figure 6.25. LFEPR field cycled signal from kaolin clay with 14% Fe_2O_3 after firing at 500 °C showing the ferro/ferrimagnetic property of the sample. A wide-line low frequency electron paramagnetic resonance spectrometer. Reproduced from [50] with permission. © 2017 Wiley Periodicals, Inc.

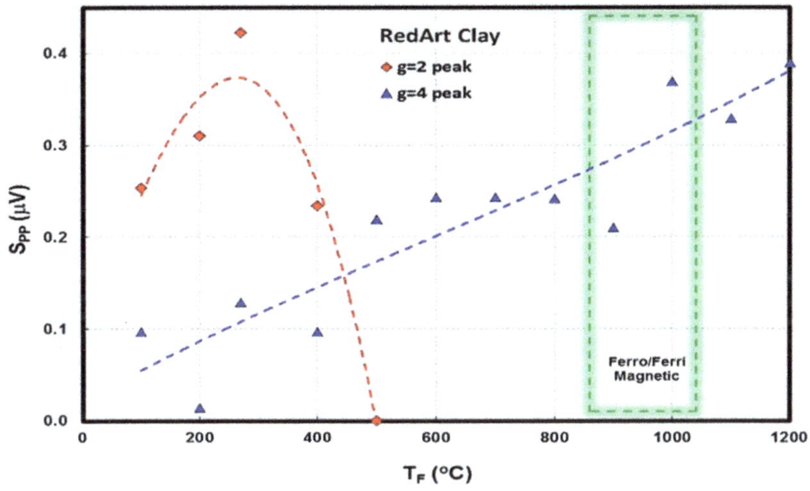

Figure 6.26. S_{PP} of the major peaks in red art clay as a function of firing temperature. Dashed lines are drawn to guide the eye.

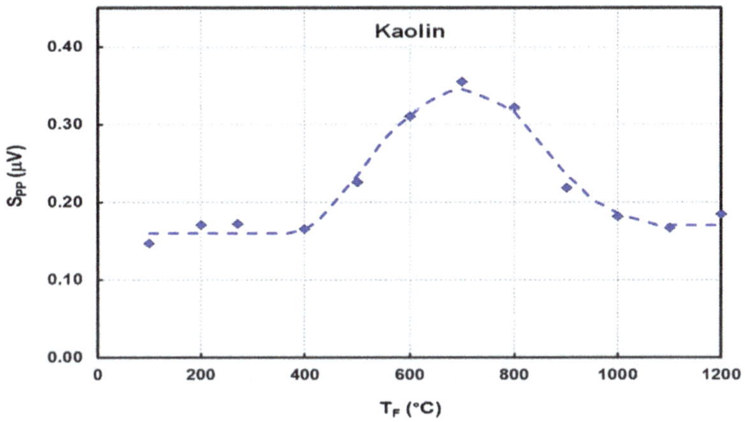

Figure 6.27. S_{PP} of the major peak in kaolin clay as a function of firing temperature. Dashed line is drawn to guide the eye.

The LFEPR spectra of kaolin clay at all firing temperatures show only one $g \approx 4$ peak. The S_{PP} versus firing temperature for kaolin clay is presented in figure 6.27. The signal is constant at low temperatures, increases to a maximum at approximately 700 °C, and then decreases to approximately the same low temperature value by 1100 °C. It would appear as though the iron in the two clays are in different environments. This observation can be confirmed by comparing the S_{PP} vs T_F for these two clays to those for Fe_2O_3, and kaolin with 3.5%, 7%, and 14% Fe_2O_3 added. The LFEPR signal from Fe_2O_3 presented in figure 6.28 displays ferromagnetic properties between 20 °C and 500 °C and at 1200 °C. There is a $g \approx 4$ paramagnetic

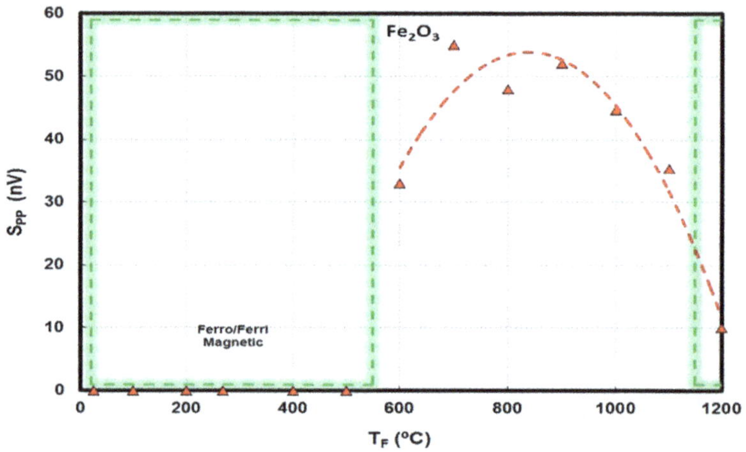

Figure 6.28. S_{PP} of Fe_2O_3 as a function of firing temperature. Dashed line is drawn to guide the eye.

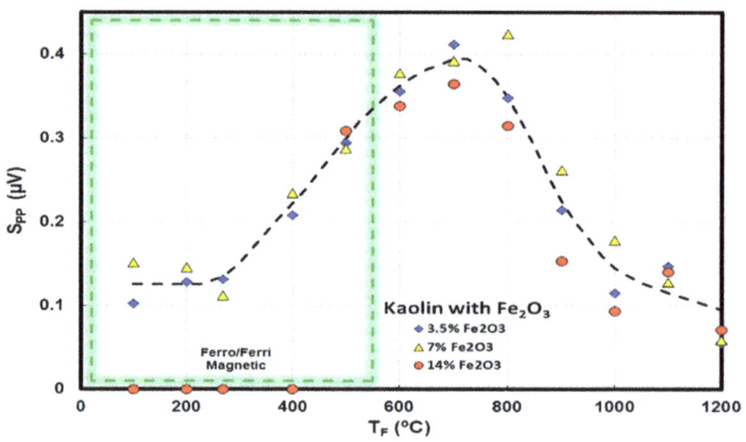

Figure 6.29. S_{PP} of the major peaks in kaolin clay and various added percentages of Fe_2O_3 as a function of firing temperature. Dashed line is drawn to guide the eye.

resonance signal between 600 °C and 1200 °C that peaks at ~850 °C. Note that the S_{PP} for the $g \approx 4$ component is approximately one tenth of that for similar peak in kaolin.

The plots of S_{PP} vs T_F for the mixtures of kaolin and Fe_2O_3 look very similar to the plot for pure kaolin. (See figure 6.29.) There is a ferro/ferromagnetic region a low T_F similar to that for Fe_2O_3. There is a $g \approx 4$ resonance between 100 °C and 1200 °C that peaks at ~700 °C. This peak is independent of added Fe_2O_3 concentration.

All these observations support the conjecture by Jinhui, Lu, and Xiasheng that the iron in kaolin is in the clay lattice [100]. Mixing Fe_2O_3 with kaolin may give the clay the same appearance as red art clay, but the lattice structure is different. These results show that LFEPR could be used to study the firing temperature of red art clay, kaolin clay, and mixtures of Fe_2O_3 and kaolin clay. LFEPR could also distinguish between red art and mixtures of kaolin and Fe_2O_3 designed to look like red art.

6.4.2.2 Ceramic objects

Five ceramic objects were studied to ascertain the possibility of using LFEPR to study intact ceramic objects. Table 6.12 lists the six ceramic objects studied, their size, and approximate age. All LFEPR signals were from Fe(III), however some were paramagnetic and others were ferro/ferrimagnetic.

The flower pot (figure 6.30) is a modern era, earthenware ceramic object. These ubiquitous objects are made of terracotta red clay and fired at a temperature of at least 1010 °C. There is great compositional variation between the terracotta clays used to make this type of earthenware. These impurities, compared to a kaolin clay, cause this earthenware to fire harder at a lower temperature. Terracotta earthenware flower pots come in a variety of sizes, making it easy to find one which will fit in the benchtop magnet and studied with either the 2.9 mm or 2 cm surface coils, or using

Table 6.12. The six ceramic objects studied by LFEPR.

Object	Description	Date	Size (cm)	Signal
Flower pot	Terracotta clay	2010	w/h = 11/10	Fe(III)
Coin	Saxon Notgeld	1920	w/h = 2.2/0.2	Fe(III)
Jug	Wedgwood Rosso Antico (ferruginous illitic clay)	1850	w/h = 12.7/14.6	Fe(III)
Candlestick	Meissen contemporary Böttger red stoneware	1920	w/h = 6/12.7	Fe(III)
Teapot	Chinese Jixing stoneware	1720	w/l/h = 10/20/13	Fe(III)
Bowl	Ming dynasty porcelain (kaolinite with iron impurities)	1500	w/h = 15.3/7	Fe(III)

Figure 6.30. Image of the terracotta red clay flower pot.

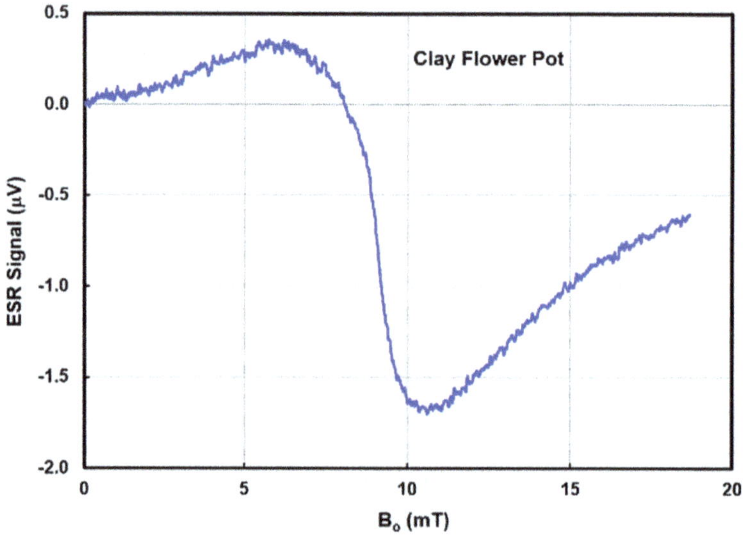

Figure 6.31. LFEPR spectrum of the terracotta red clay earthenware flower pot.

Figure 6.32. Image of top and bottom of the Notgeld coin.

the EPR MOUSE. Figure 6.31 presents the 270 MHz LFEPR spectrum recorded using the 2 cm surface coil.

Figure 6.32 displays a 1921, 50 Pfennig, 2.1 cm diameter, reddish brown, porcelain stoneware, Notgeld coin minted by the Meissen porcelain works in Saxony, Germany. The coin displays the trademark crossed swords of the Meissen porcelain manufactory. The EPR spectrum appears at first inspection like a partial first derivative of an absorption peak from a paramagnetic resonance signal with a zero crossing at 6 mT at 249.6 MHz (figure 6.33). In reality it is not, as the sample is ferro/ferrimagnetic. Figure 6.34 displays the field cycled signal from the sample using the 2.9 mm surface coil. It is clear from this figure that the sample is ferro/ferrimagnetic and how a simple single scan can be misleading.

The teapot presented in figure 6.35 is a piece of Chinese Jixing Stoneware from 1720. Spectra were taken at 272.14 MHz with the 2.9 mm diameter surface coil.

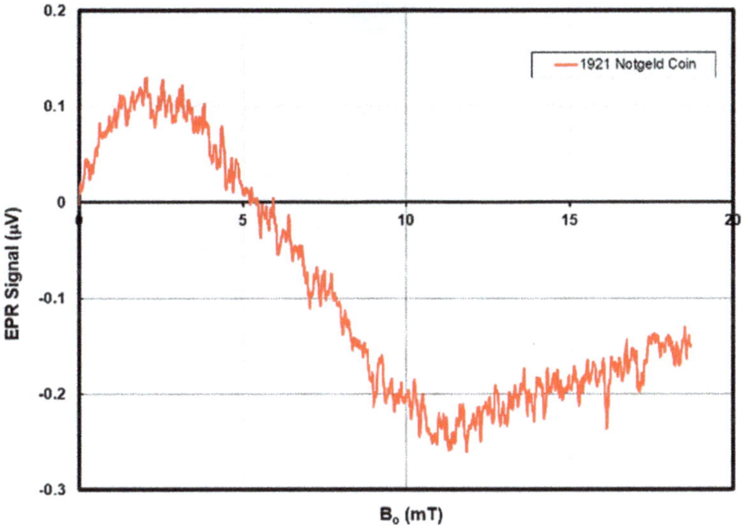

Figure 6.33. 249.6 MHz LFEPR spectrum of the Notgeld coil.

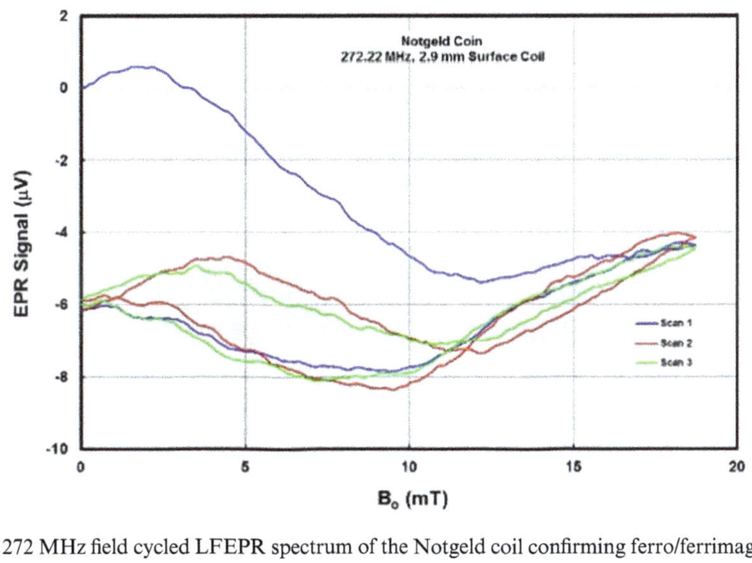

Figure 6.34. 272 MHz field cycled LFEPR spectrum of the Notgeld coil confirming ferro/ferrimagnetism in the sample.

Neither the top nor the bottom had a paramagnetic resonance signal. Both the top and bottom were however ferro/ferrimagnetic. Figure 6.36 displays the field cycled EPR signal from the spectrometer for the top and bottom of the teapot. Although not identical, the traces from the top and bottom had similar features.

Figure 6.37 displays images of the next three objects. The first is the Wedgwood Rosso Antico jug made from ferruginous illitic clay. The 421.7 MHz LFEPR

Figure 6.35. A Chinese Jixing stoneware teapot from 1720.

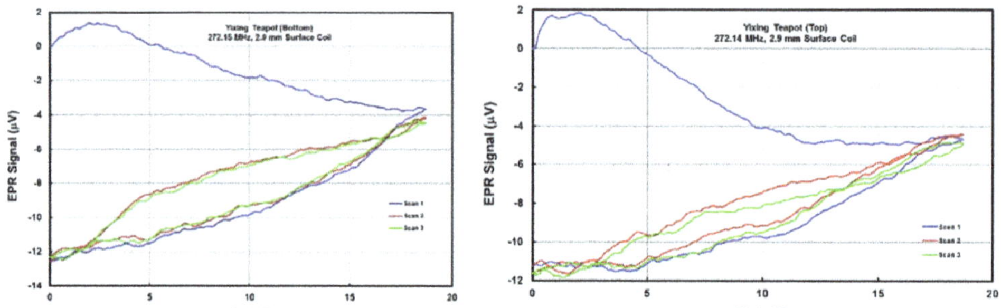

Figure 6.36. 272 MHz field cycled LFEPR spectrum of the bottom(left) and top (right) of the teapot confirming ferro/ferrimagnetism in the sample.

Figure 6.37. Image of the Wedgwood Rosso Antico jug, Meissen contemporary Böttger red stoneware candlestick, and Ming dynasty bowl.

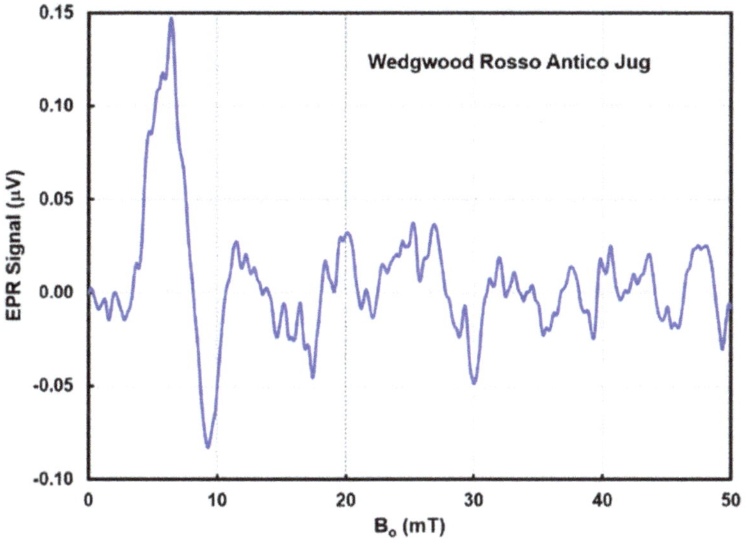

Figure 6.38. 421.7 MHz LFEPR spectrum of the Wedgwood Rosso Antico jug. Spectrum is the result of nine averages. A wide-line low frequency electron paramagnetic resonance spectrometer. Reproduced from [50] with permission. © 2017 Wiley Periodicals, Inc.

spectrum of the jug is presented in figure 6.38. This spectrum is the result of nine averages and was baseline corrected to enable the $g \approx 4$ peak to be more discernable. It is notable how small this signal is and that the spectrum is devoid of ferro/ferrimagnetism. If the clay is a kaolin like clay with Fe_2O_3 added for color, the firing temperature study would suggest that the jug is fired slightly below 1200 °C where the signal is tending towards zero. If it was fired above 1200 °C, the sample could develop ferro/ferrimagnetism.

The Meissen contemporary Böttger red stoneware candlestick dating from 1920 is pictured in figure 6.37. The candlestick did not have a paramagnetic signal but was ferro/ferrimagnetic. See figure 6.39 for the field cycled signal from the LFEPR spectrometer. Both the candlestick and the Notgeld coil originate from the Meissen factory around the same time. It is not surprising that their field cycled spectra look very similar.

The Chinese Ming dynasty bowl dating from 1500 is pictured in figures 6.37 and 6.40. The bowl is an example of the maximum size object that can fit in the solenoidal magnet given that it must also fit inside the 15 cm diameter Helmholtz style modulation coils. This porcelain bowl is made from kaolinite. Iron impurities in the kaolinite give the bowl the LFEPR spectrum presented in figure 6.41. This spectrum was recorded using the two-inch spiral surface coil as depicted in figure 6.40.

All the ceramic pieces contained iron (III), and some were ferro/ferromagnetic and others were paramagnetic. The candlestick, coin, teapot, and jug were made in a period when red stoneware was popular, even though a similar product could be made without the reddish brown color. The reddish brown color was achieved by the

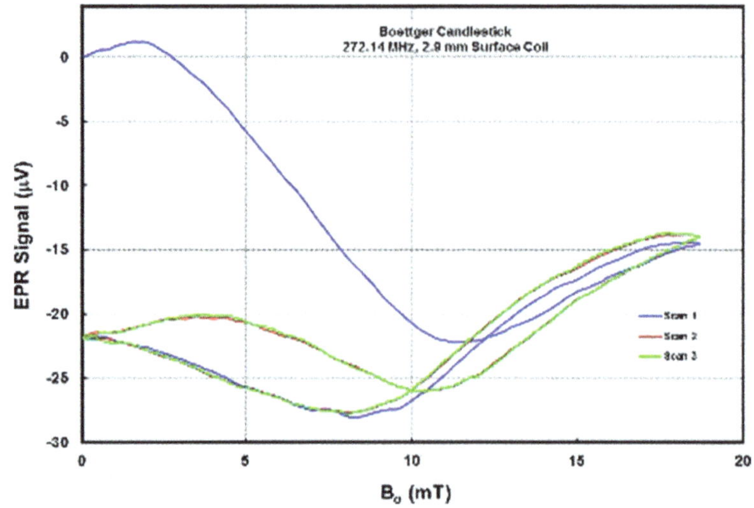

Figure 6.39. 272 MHz field cycled LFEPR spectrum of the Böttger red stoneware candlestick confirming ferro/ ferrimagnetism in the sample.

Figure 6.40. Image of the Ming dynasty bowl in the Helmholtz modulation coil and within the solenoidal magnet. The 2 cm diameter spiral surface coil is resting on the bottom of the bowl. A wide-line low frequency electron paramagnetic resonance spectrometer. Reproduced from [50] with permission. © 2017 Wiley Periodicals, Inc.

addition of iron oxide. If these ceramic pieces behave similar to the kaolin with Fe_2O_3 mixtures in the temperature section study, it is reasonable to extrapolate that there may be three firing temperature windows with different magnetic properties. In the case of the kaolin + Fe_2O_3 study these were below 550 °C where the ceramic is ferro/ferromagnetic, one with 550 °C $< T_F <$ 1150 °C where the material is paramagnetic, and one at approximately 1200 °C where the sample is again ferro/ ferromagnetic. If this hypothesis is correct, the teapot, coin, and candlestick were fired at slightly higher temperatures, and the jug at a slightly lower temperature.

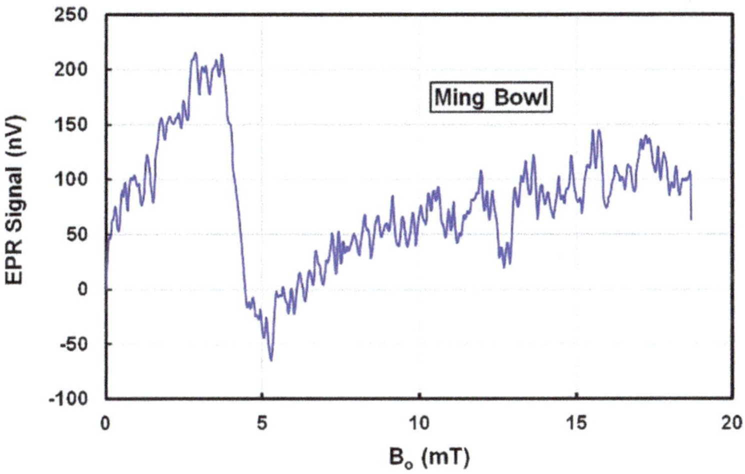

Figure 6.41. LFEPR spectrum of the Ming dynasty bowl. A wide-line low frequency electron paramagnetic resonance spectrometer. Reproduced from [50] with permission. © 2017 Wiley Periodicals, Inc.

The word slightly is used because all objects had to be fired at similar temperatures to achieve stoneware.

The bowl and flower pot were fired in the paramagnetic window. It is not unreasonable to think that the flower pot was fired at the lowest temperatures they are mass produced to be inexpensive and disposable. It is interesting to note that the flower pot has a slightly different spectrum than that of the bowl.

6.4.3 Marble

Marble possesses an EPR signal from trace amounts of Mn(II) substituted for Ca (II) in the calcium carbonate lattice of calcite or calcium magnesium carbonate lattice of dolomite. Manganese has an electron spin of 5/2 and a nuclear spin of 5/2. The energy level diagram of Mn(II) was presented earlier in figure 6.2. A high frequency EPR can detect the dominant sextet pattern of Mn(II) with a total of 22 peaks resolvable in high resolution situations. As the EPR frequency is decreased, many more transitions are observable due to the converging energy levels and zero-field energy [60]. Crystallization of the marble during the metamorphic process of converting calcite and dolomite into marble creates an orientation dependent signal that can be used to identify the origin of the marble [45]. It is worth recalling that the EPR spectrum of manganese carbonate (rhodochrosite) at 271 MHz is a single 32.1 mT homogeneously broadened absorption due to dipolar coupling between the manganese ions [69]. Figure 6.42 presents the complexity of the LFEPR spectra of a piece of marble as a function of ν between $0 < B_o < 35$ μT. Spectra were recorded at ν = 150, 210, 250, 298, 355, 410, and 440 MHz. The diagonal lines represent constant *g*-factors. Note that the spectral peaks do not follow these lines. It is hoped that the orientation dependent signal can be used to identify the origin of the marble.

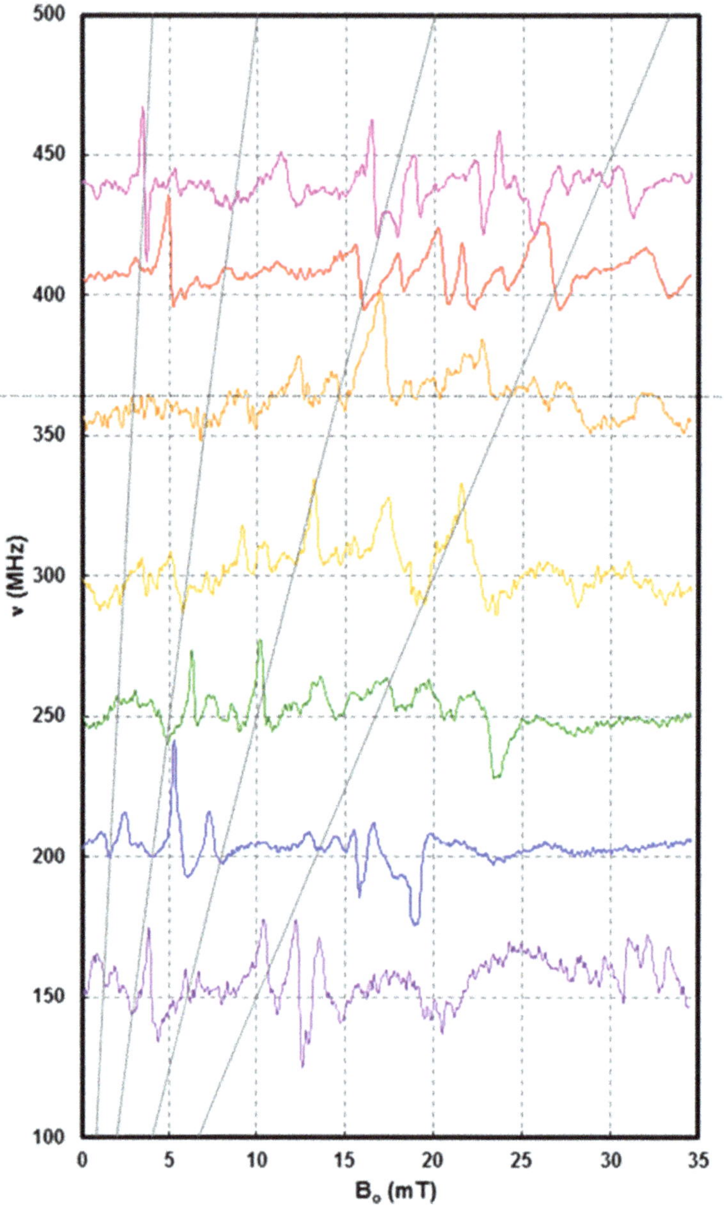

Figure 6.42. LFEPR spectra of a piece of marble recorded at seven different frequencies. Diagonal lines represent constant *g*-factor values.

6.5 Future

The future looks promising for LFEPR spectroscopy of objects with cultural heritage significance. LFEPR spectroscopy's non-invasive nature opens up the possibility of performing EPR spectroscopy on samples too precious to examine invasively. LFEPR spectroscopy will never be able to examine all objects, but neither can the other more established UV–vis, IR, Raman, XRPD, XRF, PIXE, Mössbauer, mass, NAA, and NMR spectroscopic techniques currently in use. LFEPR will always be restricted in paramagnetic, ferromagnetic, and ferrimagnetic materials. LFEPR spectroscopy will complement the other techniques used to study objects with cultural heritage significance and provide bulk rather than surface information.

The greatest challenge to be overcome in LFEPR spectroscopy is its limit of detection imposed by Boltzmann statistics. Focusing effort on a higher power RF bridge, improved bridge detection electronics, a higher modulation frequency to overcome RF source phase noise, and a more homogeneous MOUSE magnetic field should improve the limit of detection. Increasing the magnetic field sweep should open up the possibility of examining additional transition metal pigments.

In the future, we hope to make the EPR MOUSE more portable by duplicate the components of the LFEPR that it is currently tethered to. The entire unit that the EPR MOUSE will be connected to should be no more than eight inches of a 19 inch rack mount chassis.

Acknowledgements

I thank the many colleagues who have worked with me on developing hardware and applications for the LFEPR. These included Dr Nicholas Zumbulyadis, William J Ryan, Lauren E Switala, Elizabeth A Bogart, Haley Wiskoski, Matina Chanthavongsay, Akul Gupta, Merlin Hoffman, Stephany Javier Santana, Baron E Black, Celia A Mercovich, Anjana Seshadri, Emma I Hornak, and Wyatt Brown.

List of symbols

AFC	Automatic frequency control
A/D	Analog-to-digital converter
B_{Mod}	Maximum modulation magnetic field
B_o	Applied magnetic field
$B_{o\mathrm{Max}}$	Ending value for a magnetic field sweep
$B_{o\mathrm{Min}}$	Starting value for a magnetic field sweep
B_S	Saturation magnetization
B_R	Remanent magnetization
BV	Blue vitriol
C	Capacitance
C	Charcoal
CPVC	Critical pigment volume concentration
ΔE	Energy difference
dB	Decibel
D/A	Digital-to-analog converter
DBM	Doubly balanced mixer
DPPH	2,2-diphenyl-1-picrylhydrazyl
EB	Egyptian blue
EDXRF	Energy dispersive x-ray fluorescence
EPR	Electron paramagnetic resonance
f_{AFC}	AFC frequency

f_{MOD}	Magnetic field modulation frequency
FTIR	Fourier transform infrared
g	Lande' g-factor
h	Planck's constant
h	Height
H	Applied magnetic field intensity
H_C	Coercivity
H_{Ci}	Intrinsic coercivity
HB	Han blue
ID	Inside diameter
k	Boltzmann's constant
l	Length
L	Inductance
LFEPR	Low frequency electron paramagnetic resonance
M	Magnetization
I	Maxim value of the nuclear spin quantum number
M_I	Nuclear spin quantum number
S	Maxim value of the electron spin quantum number
S_{PP}	Peak-to-peak signal
M_S	Electron spin quantum number
MRI	Magnetic resonance imaging
MOUSE	Mobile Universal Surface Explorer
N^+	Number of spins in the+spin state
N^-	Number of spins in the—spin state
NMR	Nuclear magnetic resonance
OD	Outside diameter
PVC	Pigment volume concentration
Q	Resonant circuit quality factor
RF	Radio frequency
RC	Rhodochrosite
RS	Raman spectroscopy
SC	Surface coil
SERS	Surface enhanced Raman spectroscopy
SERRS	Surface enhanced resonance Raman spectroscopy
SSNMR	Solid state nuclear magnetic resonance
STS	Single turn solenoids
t	Time
t_{Wait}	Wait time
T	Temperature
T_1	Spin-lattice relaxation time
T_2	Spin–spin relaxation time
TCR	Terracotta red
t_{DW}	Dwell time
T_F	Firing temperature
UM	Ultramarine
UV	Ultraviolet
V	Volts
V_E	AFC error voltage
V_T	AFC threshold voltage
Vis	Visible

w	width
WDXRF	Wavelength dispersive x-ray fluorescence
XRF	X-ray fluorescence
β	Bohr magneton
$\delta\delta\delta B_o$	Magnetic field sweep step size
$\Delta\phi$	Lock-in phase step for LabVIEW program
ϕ_{Min}	Start phase
ϕ_{Max}	Stop phase
Ω	Impedance
μ	Permeability
Γ_{PP}	Peak-to-peak linewidth
ν	Resonance frequency
ν_P	Resonance frequency of probe

References

[1] Delaney J K, Thoury M, Zeibel J G, Ricciardi P, Morales K M and Dooley K A 2016 Visible and infrared imaging spectroscopy of paintings and improved reflectography *Herit. Sci.* **4** 1–10

[2] Delaney J K, Zeibelb J G, Thourya M, Littetonb R, Moralesa K M, Palmera M and de la Rie E R 2009 Visible and infrared reflectance imaging spectroscopy of paintings: pigment mapping and improved infrared reflectography *Proc. SPIE* **7391** 739103

[3] Piqué F and Verri G (ed) 2015 *Organic Materials in Wall Paintings* (Los Angeles, CA: The Getty Conservation Institute) p 137

[4] Giorgi L, Nevin A, Comelli D, Frizzi T, Alberti R, Zendri E, Piccolo M and Izzo F 2019 In-situ technical study of modern paintings - Part 2: Imaging and spectroscopic analysis of zinc white in paintings from 1889 to 1940 by Alessandro Milesi (1856-1945) *Spectrochim. Acta* A **219** 504–8

[5] Gabrieli F, Dooley K A, Facini M and Delaney J K 2019 Near-UV to mid-IR reflectance imaging spectroscopy of paintings on the microscale *Sci. Adv.* **5** eaaw7794

[6] Delaney J K, Zeibelb J G, Thourya M, Littetonb R, Moralesa K M, Palmera M and de la Rie E R 2009 Visible and infrared reflectance imaging spectroscopy of paintings: pigment mapping and improved infrared reflectography *Proc. SPIE* **7391** 739103

[7] Liang H, Peric B, Hughes M, Podoleanub A, Spring M and Saunders D 2007 Optical coherence tomography for art conservation & archaeology *Proc. SPIE* **6618** 661805

[8] Rosi F, Cartechini L, Sali D and Miliani C 2019 Recent trends in the application of Fourier Transform Infrared (FT-IR) spectroscopy in Heritage Science: from micro- to non-invasive FT-IR *Phys. Sci. Rev.* **4** 20180006

[9] Salvadó N, Butí S, Tobin M J, Pantos E, Prag A J N W and Pradell T 2005 Advantages of the use of SR-FT-IR microspectroscopy: applications to cultural heritage *Anal. Chem.* **77** 3444–51

[10] Henderson E J, Helwig K, Read S and Rosendahl S M 2019 Infrared chemical mapping of degradation products in cross-sections from paintings and painted objects *Herit. Sci.* **7** 71

[11] Buti D, Rosi F, Brunetti B G and Miliani C 2013 In-situ identification of copper-based green pigments on paintings and manuscripts by reflection FTIR *Anal. Bioanal. Chem.* **405** 2699–711

[12] Rosi F, Grazia C, Gabrieli F, Romani A, Paolantoni M, Vivani R, Brunetti B G, Colomban P and Miliani C 2016 UV–Vis-NIR and micro Raman spectroscopies for the non destructive identification of $Cd_{1-x}Zn_xS$ solid solutions in cadmium yellow pigments *Microchem. J.* **124** 856–67

[13] Brosseau C L, Rayner K S, Casadio F, Grzywacz C M and Van Duyne R P 2009 Surface-enhanced Raman spectroscopy: a direct method to identify colorants in various artist media *Anal. Chem.* **81** 7443–7

[14] Cottingham K 2009 SERS enters the art scene *Anal. Chem.* **81** 7128

[15] Leona M 2009 Microanalysis of organic pigments and glazes in polychrome works of art by surface-enhanced resonance Raman scattering *Proc. Natl. Acad. Sci.* **106** 14757–62

[16] De Meyer S *et al* 2019 Macroscopic x-ray powder diffraction imaging reveals Vermeer's discriminating use of lead white pigments in Girl with a Pearl Earring *Sci. Adv.* **5** eaax1975

[17] Janssens K, Dik J, Cotte M and Susini J 2010 Photon-based techniques for nondestructive subsurface analysis of painted cultural heritage artifacts *Acc. Chem. Res.* **43** 814–25

[18] Janssens K *et al* 2013 Non-invasive and non-destructive examination of artistic pigments, paints, and paintings by means of x-ray methods *Top. Curr. Chem.* **374** 81

[19] Kanngiesser B, Mantouvalou I and Malzer W 2008 Non-destructive, depth resolved investigation of corrosion layers of historical glass objects by 3D micro x-ray fluorescence analysis *J. Anal. At. Spectrom.* **23** 814–9

[20] Prati S, Fuentes D, Sciutto G and Mazzeo R 2014 The use of laser photolysis-GC-MS for the analysis of paint cross sections *J. Anal. Appl. Pyrolysis* **105** 327–34

[21] Colombini M P, Andreotti A, Lanterna G and Rizzi M 2003 A novel approach for high selective micro-sampling of organic painting materials by Er:YAG laser ablation *J. Cult. Herit.* **4** 355s–61s

[22] Russell J, Singer B W, Perry J J and Bacon A 2011 The identification of synthetic organic pigments in modern paints and modern paintings using pyrolysis-gas chromatography–mass spectrometry *Anal. Bioanal. Chem.* **400** 1473–91

[23] Cotter M J 1981 Neutron Activation Analysis of Paintings: Autoradiography following neutron irradiation provides information for authentication and restoration and reveals the existence of underpaintings undetected by other means *Am. Sci.* **69** 17–27

[24] Calligaro T, Gonzalez V and Pichon L 2015 PIXE analysis of historical paintings: Is the gain worth the risk? *Nucl. Instrum. Methods* B **363** 135–43

[25] Rizzutto M A, Moro M V, Silva T F, Trindade G F, Added N, Tabacniks M H, Kajiya E M, Campos P H V, Magalhães A G and Barbosa B 2014 External-PIXE analysis for the study of pigments from a painting from the Museum of Contemporary Art *Nucl. Instrum. Methods* B **332** 411–4

[26] Lehmann R *et al* 2017 artwork: red-pigment material investigation by backscattering Fe-57 Mössbauer spectroscopy, SEM and p-XRF *STAR* **3** 450–5

[27] Spyros A and Anglos D 2004 Study of aging in oil paintings by 1D and 2D NMR spectroscopy *Anal. Chem.* **76** 4929–36

[28] Proietti N, Capitani D and Di Tullio V 2018 Nuclear magnetic resonance, a powerful tool in cultural heritage *Magnetochemistry* **4** 11

[29] Kehlet C, Kuvvetli F, Catalano A and Dittmer J 2016 Solid-state NMR for the study of Asger Jorn's paintings *Microchem. J.* **125** 308–14

[30] Catalano J, Murphy A, Yao Y, Zumbulyadis N, Centeno S and Dymbowski C 2019 *Understanding the Dynamics and Structure of Lead Soaps in Oil Paintings using*

Multinuclear NMR ed F Casadio, K Keune, P Noble, A Van Loon, E Hendriks, S A Centeno and G Osmond (Switzerland: Springer) pp 69–84

[31] Blümich B and Baias M 2017 Nondestructive testing of objects from cultural heritage with NMR *Modern Magnetic Resonance* ed G Webb (Cham: Springer)

[32] Presciutti F, Perlo J, Casanova F, Glöggler S, Milian C, Blümich B, Brunetti B G and Sgamellotti A 2008 Noninvasive nuclear magnetic resonance profiling of painting layers *Appl. Phys. Lett.* **93** 033505

[33] Le Pape L 2016 Applications of EPR in studies of archaeological samples *Modern Magnetic Resoance* (Cham: Springer) pp 1–25

[34] Christiansen M B, Sørensen M A, Sanyova J, Bendix J and Simonsen K P 2017 Characterisation of the rare cadmium chromate pigment in a 19th century tube colour by Raman, FTIR, X-ray and EPR *Spectrochim. Acta Part* A **175** 208–14

[35] Gobeltz N, Demortier A, Lelieur J P and Duhayon C 1998 Correlation between EPR, Raman and colorimetric characteristics of the blue ultramarine pigments *J. Chem. Soc. Faraday Trans.* **94** 677–81

[36] Lakshmi Reddy S, Udayabashakar Reddy G, Ravindra Reddy T, Thomas A R, Rama Subba Reddy R, Frost R L and Endo T 2015 XRD, TEM, EPR, IR and nonlinear optical studies of yellow ochre *J. Laser Opt. Photon.* **2** 120

[37] Orsega E F, Agnoli F and Mazzocchin G A 2006 An EPR study on ancient and newly synthesised Egyptian blue *Talenta* **68** 831–5

[38] Raulin K, Gobeltz N, Vezin H, Touati N, Lede B and Moissette A 2011 Identification of the EPR signal of $S2^{-}$ in green ultramarine pigments *Phys. Chem. Chem. Phys.* **13** 9253–9

[39] Bensimon Y, Deroide B, Clavel S and Zanchetta J-V 1998 Electron spin resonance and dilatometric studies of ancient ceramics applied to the determination of firing temperature *Jpn. J. Appl. Phys.* **37** 4367

[40] Presciutti F, Capitani D, Sgamellotti A, Brunetti B G, Costantino F, Viel S and Segre A 2005 Electron paramagnetic resonance, scanning electron microscopy with energy dispersion x-ray spectrometry, x-ray powder diffraction, and NMR characterization of iron-rich fired clays *J. Phys. Chem.* B **109** 22147–58

[41] Warashina T, Higashimura T and Maeda Y 1981 Determination of the firing temperature of ancient pottery by means of ESR spectrometry *Br. Mus. Occas.* **19** 117–28

[42] Dietemann P, Kalin M, Zumbuhl S, Knochenmuss R, Wulfert S and Zenobi R 2001 A mass spectrometry and electron paramagnetic resonance study of photochemical and thermal aging of triterpenoid varnishes *Anal. Chem.* **73** 2087–96

[43] Polikreti K, Maniatis Y, Bassiakos Y, Kourou N and Karageorghis V 2004 Provenance of archaeological limestone with EPR spectroscopy: the case of the Cypriote-type statuettes *J. Archaeol. Sci.* **31** 1015–28

[44] Polikreti K and Maniatis Y 2004 Distribution of changes of Mn(II) and Fe(III) on weathered marble surfaces measured by EPR spectroscopy *Atmos. Environ.* **38** 3617

[45] Polikreti K and Maniatis Y 2002 A new method for the provenance of marble based on EPR spectroscopy *Archaeometry* **44** 1–21

[46] Attanasio D, Capitani D, Federici C and Segre A L 1995 Electron spin resonance study of paper samples dating from the fifteenth to the eighteenth century *Archaeometry* **37** 377–84

[47] Moretto L M, Orsega E F and Mazzocchin G A 2011 Spectroscopic methods for the analysis of celadonite and glauconite in Roman green wall paintings *J. Cult. Herit.* **12** 384–91

[48] Rowe M W 2010 Dating of rock paintings in the Americas: a word of caution *Proc. of the Int. Federation of Rock Art Congress* pp 573–84

[49] Switala L E, Black B E, Mercovich C A, Seshadri A and Hornak J P 2017 An electron paramagnetic resonance mobile universal surface explorer *J. Magn. Reson.* **285** 18–25

[50] Switala L E, Ryan W J, Hoffman M, Javier Santana S, Black B E and Hornak J P 2017 A wide-line low frequency electron paramagnetic resonance spectrometer *Concepts Magn. Reson.* **47B** e21355

[51] Lee S J, Tsai C C, Cho H, Seo M, Eom T, Nam W, Lee Y P and Ketterson J B 2009 Hysteretic characteristics of low-field microwave absorption of a Co thin film *J. Appl. Phys.* **106** 063922

[52] Hornak J P, Spacher M and Bryant R G 1991 A modular low frequency ESR spectrometer *Meas. Sci. Technol.* **2** 520–2

[53] Szczepaniak E and Hornak J P 1993 ESR imaging based on the modulation-field phase *J. Magn. Reson.* A **104** 315–20

[54] Salikhov I, Walczak T, Lesniewski P, Khan N, Iwasaki A, Comi R, Buckey J and Swartz H M 2005 EPR spectrometer for clinical applications *Magn. Reson. Med.* **54** 1317–20

[55] Rinard G A, Quine R W, Eaton G R, Eaton S S, Barth E D, Pelizzari C A and Halpern H J 2002 Magnet and gradient coil system for low-field EPR imaging *Concepts Magn. Reson.* **15** 51–8

[56] Zumbulyadis N 2006 *Meissen's Blue And White Porcelain: Dining in Royal Splendor* (Atglen, PA: Schiffer)

[57] Zumbulyadis N 2010 Böttger's Eureka! New insights into the European reinvention of porcelain *Bull. Hist. Chem.* **35** 24–32

[58] Zumbulyadis N 2013 '...with a dreadful bang'—a chemical history of Böttger Lustre *Keramos* **222** 3–16

[59] Poole C P Jr 1983 *Electron Spin Resonance* 2nd edn (New York: Wiley) p 780

[60] Piligkos S, Laursen I, Morgenstjerne A and Weihe H 2007 Sign and magnitude of spin Hamiltonian parameters for Mn^{2+} impurities in calcite. A multi- and low-frequency EPR study *Mol. Phys.* **105** 2025–30

[61] Lobanov N N *et al* 2009 Differentiation of magnetic composites in terms of their nanostructural organization *Dokl. Chem.* **426** 96–100

[62] Hyde J S and Gajdzinski J 1988 EPR automatic frequency control circuit with field effect transistor (FET) microwave amplification *Rev. Sci. Instrum.* **59** 159–67

[63] Hirata H and Luo Z-W 2001 Stability analysis and design of automatic frequency control system for *in vivo* EPR spectroscopy *Magn. Reson. Med.* **46** 1209–15

[64] Hirata H and Hirotada F 2007 Automatic matching control circuit based on phase sensitive detection for *in vivo* CW-EPR spectroscopy *Meas. Sci. Technol.* **18** N27–31

[65] Quine R W, Rinard G A, Eaton S S and Eaton G R 2002 A pulsed and continuous wave 250 MHz electron paramagnetic resonance spectrometer *Concepts Magn. Reson.* **15** 59–91

[66] Koscielniak J, Devasahayam N, Moni M S, Kuppusamy P, Yamada K, Mitchell J B, Krishna M C and Subramanian S 2000 300 MHz continuous wave electron paramagnetic resonance spectrometer for small animal *in vivo* imaging *Rev. Sci. Instrum.* **71** 4273–81

[67] Meeker D 2019 Finite Element Method Magnetics www.femm.info/wiki/HomePage (Accessed 10 March 2020)

[68] Troup G J, Hutton D R, Bacci M, Lotti F, Casini A and Picollo M 1996 The possible use of EPR spectroscopy for paint pigment identification *Proc. of 20th ANZIP Annual Condensed Matter Physics Meeting (Wagga Wagga, NSW, Australia)* p 158

[69] Lea A S, Hurt T T, El-Azab A, Amonette J E and Baer D R 2003 Heteroexpitaxial frowth of a manganese carbonate secondary nano-phase on the (1014) surface of calcite in solution *Surf. Sci.* **524** 63–7

[70] Eaton G R *et al* 2001 Multifrequency electron paramagnetic resonance of ultramarine blue *Appl. Magn. Reson.* **21** 563–70

[71] Billing D E, Hathaway B J and Nicholls P 1969 Polarisations of the copper(II) and uranyl electronic spectra of meta-zeunerite single crystals *J. Chem. Soc.* A **1969** 316–9

[72] Navas A S, Reddya B J and Nieto F 2004 Spectroscopic study of chromium, iron, OH, fluid and mineral inclusions in uvarovite and fuchsite *Spectrochim. Acta Part A* **60** 2261–8

[73] Lan M, Beghein N, Charlier N and Gallez B 2004 Carbon blacks as EPR sensors for localized measurements of tissue oxygenation *Magn. Reson. Med.* **51** 1272–8

[74] Alter M, Binet L, Touati N, Lubin-Germain N, Hô A-S L, François Mirambet F and Didier Gourier D 2019 Photochemical origin of the darkening of copper acetate and resinate pigments in historical paintings *Inorg. Chem.* **58** 13115–28

[75] Gunn M, Chottard G, Rivière E, Girerd J-J and Chottard J-C 2002 Chemical reactions between copper pigments and oleoresinous media *Stud. Conserv.* **47** 12–23

[76] Hagiwara M, Kimura S, Nishihagi N, Suzuki T, Nohara M, Takagi H and Kindo K 2010 High field magnetism of the spinel oxide $CoAl_2O_4$ *J. Low Temp. Phys.* **159** 11–4

[77] Viana G A, Lacerda R G, Freire F L and Marques F C 2008 ESR investigation of graphite-like amorphous carbon films revealing itinerant states as the ones responsible for the signal *J. Non-Cryst. Solids* **354** 2135–7

[78] Budinova T, Petrov N, Duber S and Shebestova E 1994 EPR Investigation of oxidized anthracite *Carbon* **32** 417–22

[79] Cummerow R L and Halliday D 1946 Paramagnetic losses in two manganous salts *Phys. Rev.* **70** 433

[80] Sunandana C S 1998 Techniques and applications of electron spin resonance *Bull. Mater. Sci.* **21** 1–70

[81] Mabbs F E and Porter J K 1973 Single crystal electron spin resonance of copper(II) ammonium sulphate hexahydrate *J. Inorg. Nucl. Chem.* **35** 3219–22

[82] Otto H H 2017 Crystal growth of $Cu_6(Ge,Si)_6O_{18}$ $6H_2O$ and assignment of the UV–Vis spectra in omparison to dehydrated dioptase and selected Cu(II) oxo-compounds including cuprates *World J. Condens. Matter Phys.* **7** 57–79

[83] Reddy K M, Jacob A S and Reddy B J 1986 EPR and optical spectra of Cu^{2+} in dioptase *Ferroelectr. Lett.* **6** 103–12

[84] Zoleo A, Nodari L, Rampazzo M, Piccinelli F, Russo U, Federici C and Brustolon M 2014 Characterization of pigment and binder in badly conserved illuminations of a 15th-century manuscript *Archaeometry* **56** 496–512

[85] Reddy B J and Sarma K B N 1981 Absorption spectra of Cu^{2+} in azurite *Solid State Commun.* **38** 547–9

[86] Santoro C *et al* 2014 New highlights on degradation process of verdigris from easel paintings *Appl. Phys.* A **114** 637–45

[87] Guskos N, Typek J, Maryniak M, Żolnierkiewicz G, Podsiadly M, Arabczyk W, Lendzion-Bielun Z and Narkiewicz U 2006 Effect of calcination and structural additives on the EPR spectra of nanocrystalline cobalt oxides *Mater. Sci.-Poland* **24** 1095–102

[88] Singer L S 1955 Paramagnetic resonance absorption in some Cr^{+3} complexes *J. Chem. Phys.* **23** 379–88

[89] Goldberg I B, Crowe H R and Robertson W M 1977 Determination of the composition of sodium chromite and chromic oxide by electron spin resonance spectroscopy *Anal. Chem.* **49** 962–6

[90] Khamlich S, Srinivasu V V, Nemraouil O, McCrindle R, Cingo N and Maaza M 2011 Electron spin resonance study of alpha-Cr_2O_3 and $Cr_2O_3{\cdot}nH_2O$ quasi-spherical nano-particles *Nanosci. Nanotech. Lett.* **3** 550–5

[91] Monico L, Janssens K, Cotte M, Sorace L, Vanmeert F, Brunetti B G and Miliani C 2016 Chromium speciation methods and infrared spectroscopy for studying the chemical reactivity of lead chromate-based pigments in oil medium *Microchem. J.* **124** 272–82

[92] Monico L, Janssens K, Cotte M, Romani A, Sorace L, Grazia C, Brunettia B G and Miliania C 2015 Synchrotron-based x-ray spectromicroscopy and electron paramagnetic resonance spectroscopy to investigate the redox properties of lead chromate pigments under the effect of visible light *J. Anal. At. Spectrom.* **30** 1500–10

[93] Javier S and Hornak J P 2018 A non-destructive method of identifying pigments on canvas using electron paramagnetic resonance spectroscopy *J. Am. Inst. Conserv.* **57** 73–82

[94] Asbeck W K and Van Loo M 1947 Critical pigment volume relationships *Ind. Eng. Chem.* 1470–5

[95] Kendall R 2000 *Monet by Himself* (Secaucus, NJ: Cartwell) p 327

[96] Ren J Y, Chang C Q, Fung P C, Shen J G and Chan F H 2004 Free radical EPR spectroscopy analysis using blind source separation *J. Magn. Reson.* **166** 82–91

[97] Bogart E A, Wiskoski H, Chanthavongsay M, Gupta A and Hornak J P 2020 The noninvasive analysis of paint mixtures on canvas using an EPR MOUSE *Heritage* **3** 140–51

[98] Eaton G R, Eaton S S, Barr D P and Weber R T 2010 *Quantitative EPR* (New York: Springer) p 185

[99] Ardelean I, Peteanu M, Filip S, Simon V and Györffy G 1997 EPR and magnetic susceptibility studies of iron ions in $70TeO_2$ $25B_2O_3$ 5PbO glass matrix *Solid State Commun.* **102** 341–6

[100] Jinhui L, Lu W and Xiasheng G 1998 EPR studies of the modes of occurrence of iron in kaolins from Shange, Maoming area *Acta Petrol. Mineral.* **17** P275–80

IOP Publishing

Spectroscopic and Microscopy Techniques for Archaeological and Cultural Heritage Research (Second Edition)

Ashutosh Kumar Shukla

Chapter 7

Determination by ESR of the temperature of prehistoric lithic heat treatment

Shin Toyoda

Quantitative assessment of the temperature of past lithic heat treatment is one of the important issues in archaeology in order to study the ancient technology for lithic tools. In the present chapter, a method to contribute to this issue based on physics with the electron spin resonance (ESR) signal of the E_1' center in quartz is reviewed and summarized, being focused on the fundamentals. When the heat treatment is 'recent', where the natural radiation given to the specimen after the treatment is negligible, the range of the temperature is easily determined and temperature may be quantitatively obtained if a reference source material is available. When the age of treatment is old, it is not simple but temperature may be obtained in some conditions. The method would also be useful in dating of heated lithic tools and pottery to estimate whether dating signal has been erased at the time of heat treatment.

7.1 Introduction

The firing technology to make lithic tools is one of the important issues in archaeology. Although it is established that heat treatment was frequently used among ancient human beings, it is not easy to identify the treatment quantitatively. There have been numerous investigations on this issue using macroscopic attributes such as color and luster (Collins and Fenwick 1974), and changes of mineral assemblages and chemical compositions (Delagnes *et al* 2016 and references therein). Thermoluminescence is a method to identify heat treatment quantitatively by comparing the glow curve of artificially heated material with the one for examination (Melcher and Zimmerman 1977; Pawlish and Sheppard 1983).

doi:10.1088/978-0-7503-4846-1ch7

In the present chapter, another method in physics with electron spin resonance (ESR, equivalent to EPR, electron paramagnetic resonance) is presented which observes a paramagnetic defect in crystalline quartz. As the decay and formation follows laws in physics, the estimation can be as quantitative as thermoluminescence with an additional advantage that it is specifically clear what kind of lattice defects we are observing.

7.2 The E_1' center in quartz

There are many paramagnetic lattice defects in quartz as summarized by Weil (1984). Among them, the E_1' center is one of the most fundamental and major paramagnetic lattice defects in crystalline quartz. A model proposed by Feigl *et al* (1974) is most accepted that an unpaired electron at a single silicon sp^3 orbital oriented along a bond direction into the vacancy where that silicon relaxes toward the vacancy while the other silicon, in the opposite side, with a positive charge relaxes away in to the basal plane of its three oxygen neighbors (figure 7.1) (Rudra and Fowler 1987) where a typical ESR powder spectrum observed in natural quartz is shown in figure 7.2.

Unlike other radiation induced paramagnetic centers, the intensity of the E_1' center increases on heating up to 300 °C as shown in figure 7.3 (Toyoda and Ikeya 1991), which was already found at the early stage of the research (Weeks and Nelson 1960). The increase was attributed to an electronic process in quartz that electronic holes at such as the aluminum hole center, $[AlO_4]^0$ (Nuttal and Weil 1981), are transferred on heating to neutral oxygen vacancies with two electrons (Si–Si bond), so that they recombine one of the two electrons to form the E_1' center (Jani *et al* 1983). Above 300 °C, the intensity of the E_1' center is decreased and annealed out at 450 °C (figure 7.3).

Toyoda and Ikeya (1991) heated a sample of granitic quartz at 430 °C for 15 min almost to erase the signal of the E_1' center. When the sample was subsequently

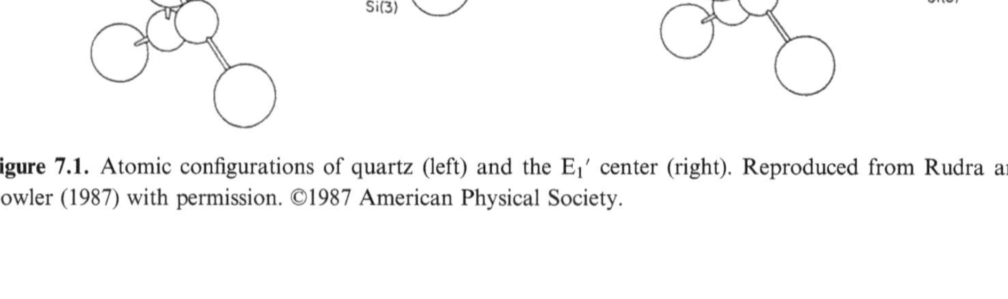

Figure 7.1. Atomic configurations of quartz (left) and the E_1' center (right). Reproduced from Rudra and Fowler (1987) with permission. ©1987 American Physical Society.

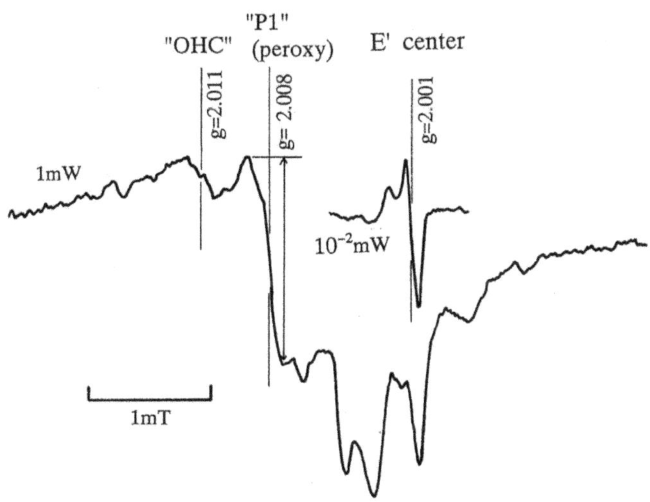

Figure 7.2. A typical ESR powder spectrum of natural quartz. Reproduced from Toyoda and Ikeya (1991).

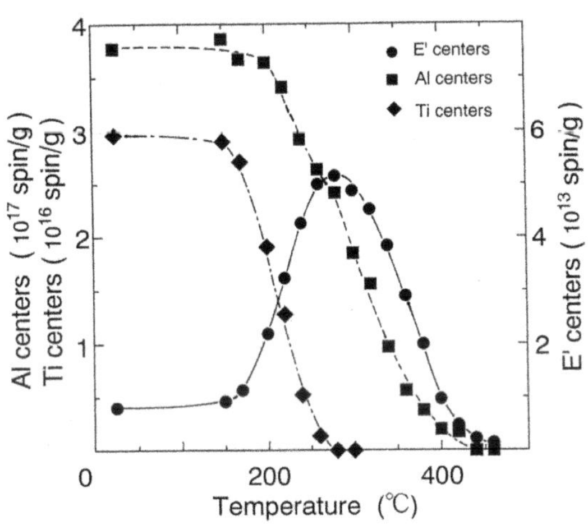

Figure 7.3. The changes of the ESR signal intensities in quartz on heating (15 min stepwise heating). Reproduced from Toyoda and Ikeya (1991).

irradiated and heated at 300 °C for 15 min, the signal of E_1' center was regenerated. The intensity has a dose response, increasing up to 200 Gy and then saturated (figure 7.4). They argued that the increase is due to the increase of the number of the holes formed by the gamma rays, to be supplied to neutral oxygen vacancies to form the E_1' center, while the saturation is due to the limitation of the number of the oxygen vacancies. Based on this argument, they proposed a protocol to estimate the number of the oxygen vacancies, which is to measure the ESR intensity of the E_1' center after gamma ray irradiation to more than 200 Gy followed by heating at 300 °C for 15 min (the heat-treated E_1' center).

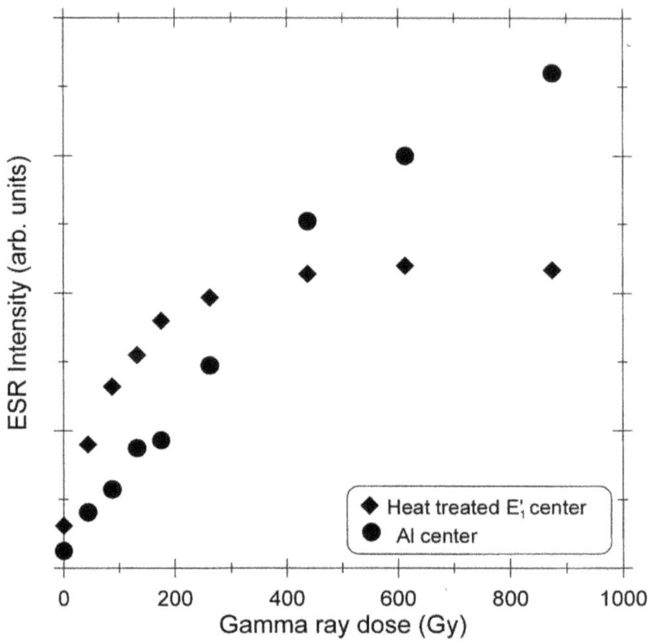

Figure 7.4. Regeneration of the E_1' center by irradiation and following heating. Reprinted from Toyoda and Hattori (2000), Copyright (2000) with permission from Elsevier. The quartz sample was first heated at 430 °C to almost erase the signal. Gamma ray doses were given as shown in the horizontal axis. The intensity of the aluminum hole center was measured, then the sample was heated at 300 °C for 15 min, and the E_1' center intensity was obtained as shown.

Toyoda and Ikeya (1991) also heated the granitic quartz between 450 °C and 600 °C following the gamma ray irradiation and second heating at 300 °C for 15 min to see the decay of the signal intensities of the regenerated E_1' center (the number of the oxygen vacancies) to obtain the result shown in figure 7.5 indicating that the oxygen vacancies decay by the first heating above 450 °C and that it is annealed out by heating above 600 °C.

Another interesting question, which is not directly related to the present application of past lithic heat treatment, is how the oxygen vacancies are created in quartz. With the above technique to estimate the number of the oxygen vacancies, Toyoda and Hattori (2000) showed that the number of the oxygen vacancies in quartz correlates with age of the host granite in the age range of 10 Ma to 1 Ga. The experimental results obtained by Toyoda *et al* (1996) indicate that the oxygen vacancies are created by gamma ray irradiation in the dose range of 10^7 Gy, Toyoda *et al* (2005) concluded that the oxygen vacancies in natural quartz are formed by beta and gamma rays at a very low production rate and accumulated in that long geologic time scale.

7.3 Estimation of the heat treatment in the 'recent' past (cases A)

Utilizing the above characteristics of the E_1' center in quartz on heating, Toyoda *et al* (1993) proposed a protocol (figure 7.6) to estimate the past lithic heat treatment

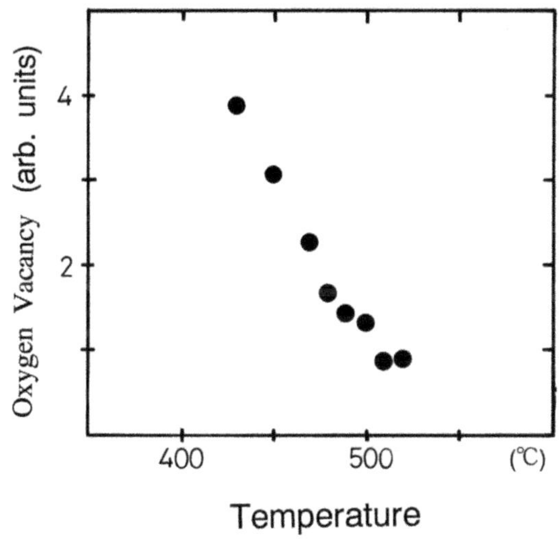

Figure 7.5. Decay of the oxygen vacancies. Reproduced from Toyoda and Ikeya (1991). The sample was first heated at the temperatures indicated in the horizontal axis. After gamma ray dose of 1 kGy was given, the samples were heated at 300 °C for 15 min and the E_1' center intensity was obtained as shown.

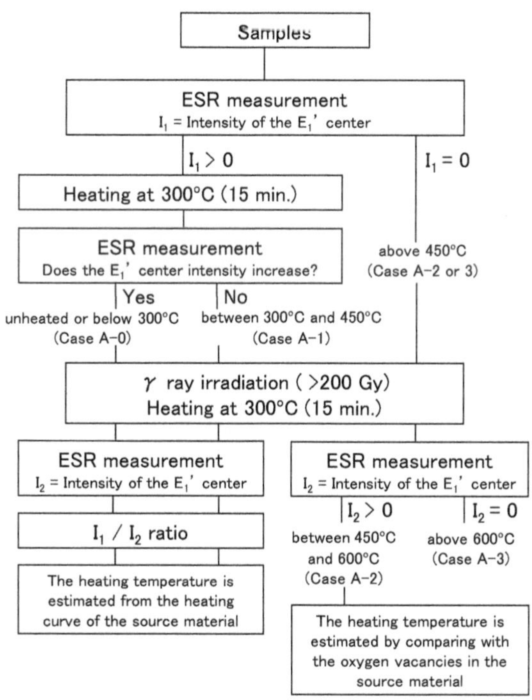

Figure 7.6. The protocol to estimate the temperature of lithic heat treatment proposed by Toyoda *et al* (1993) for 'recently' heated samples. Modified after Toyoda *et al* (1993), Copyright (1993) with permission from Elsevier.

Table 7.1. Categories defined in the present chapter.

	Temperature range (°C)			
Time of heat treatment	<300	300–450	450–600	>600
'Recent' (notations in Toyoda *et al* (1993))	A-0 (Degree 0)	A-1 (Degree 1)	A-2 (Degree 2–1)	A-3 (Degree 2–2)
About 30 ka, up to 60 ka (about 100 Gy, up to 200 Gy)	B-0	B-1	B-2	B-3
>60 ka (>200 Gy)	C-0	C-1	C-2	C-3

and showed actually that it worked for those of native Americans in 15th century. Dunnell *et al* (1994) showed that the results are consistent with the estimation that the lithic tools were heated when they were recycled. In the protocol, it is assumed that the source material is old enough to show the E_1' center (older than 10 Ma) (Toyoda and Hattori 2000) and that the heat treatment was made very recently where the natural radiation dose after the heat treatment is negligible. According to figure 7.4, the dose response is in the range of 100 Gy, therefore, the natural radiation less than 10 Gy would be considered to be negligible. In the above case of the tools of native Americans, it took only about 500 years since the heat treatment was made where the dose should be surely less than 10 Gy.

As in figure 7.6, if the E_1' center is observed in the raw specimen (the one for examination), the temperature is considered to be below 450 °C (cases A-0 or 1)* while it is above 450 °C (cases A-2 or 3) if not, see table 7.1 for definition in the present chapter. This is because the E_1' center is erased at 450 °C as shown in figure 7.3. In the former case that the E_1' center is observed, the next step is to heat the specimen at 300 °C for 15 min As the raw sample specimen is at some point in the curve in figure 7.3, if the temperature of heat treatment was below 300 °C (case A-0), the signal intensity of the E_1' center will increase on laboratory heating (300 °C for 15 min) while the intensity will not change or will make slight decrease if the heat treatment temperature was above 300 °C (case A-1).

In the latter case that the E_1' center is not observed (cases A-2 or 3), the next step is to examine whether the oxygen vacancies are annealed by heating or still present. The protocol proposed by Toyoda and Ikeya (1991) is adopted for this, i.e. gamma ray irradiation to more than 200 Gy and following heating at 300 °C for 15 min If the oxygen vacancies are present, the signal of the E_1' center will be regenerated with this procedure indicating that the heating was below 600 °C (case A-2). It is, on the other hand, estimated to be above 600 °C (case A-3) if the E_1' center is not regenerated. In that case, it is necessary to confirm that the sample contains quartz by some method such as x-ray diffraction. As for the amount of gamma ray dose to give, it should be more than 200 Gy (figure 7.4), where it is recommended to be 3 kGy in Toyoda *et al* (1993). Probably 3 kGy would be too much because it could possibly create small amount of oxygen vacancies. The author would like to recommend 1 kGy instead.

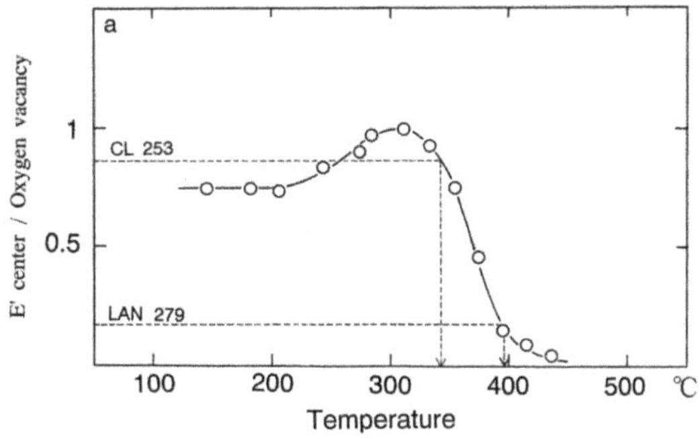

Figure 7.7. Examples of temperature estimation of lithic heat treatment with a reference source material. Reprinted from Toyoda *et al* (1993), Copyright (1993) with permission from Elsevier. The horizontal axis denotes the I_1/I_2 ratio.

When the source material of the lithic tools is available as a reference, the temperature of the heat treatment may be estimated quantitatively. Toyoda *et al* (1993) proposed to have a reference heating curve of the I_1 (the E_1' center signal intensity of the raw specimen) to I_2 (after gamma ray irradiation to 3 kGy and following heating at 300 °C for 15 min) ratio as a function of heating temperature with the reference source material. When the I_1 to I_2 ratio of the specimen is obtained, the temperature of the heat treatment is obtained as examples shown in figure 7.7 as the one corresponding to the ratio on the reference curve with considering the temperature range already obtained with the protocol shown in figure 7.6.

Although it is not discussed in Toyoda *et al* (1993), it would be possible to estimate the temperature between 450 °C and 600 °C when a reference material is available. One can obtain a reference decay curve of the oxygen vacancies like the one shown in figure 7.5 with the reference material, then, temperature may be obtained with this curve from the number of oxygen vacancies observed in the examining specimen.

The I_1 to I_2 ratios are not constant but vary from sample to sample among the natural source materials as seen when comparing figures 7.3 and 7.7. This is probably, as discussed in Toyoda and Amimoto (2020), because the ratio is in the balance between the two electronic processes in quartz. One is the natural radiation which supply electrons to convert the E_1' center to the diamagnetic oxygen vacancy with two electrons, i.e. to reduce the E_1' center signal, and the other is the thermal activation process, same as the one occurring on heating, which supply electronic holes to convert the diamagnetic oxygen vacancy to the E_1' center. Therefore, a reference material is actually needed in estimating the temperature below 300 °C. On the other hand, decays of the E_1' center and of the oxygen vacancies may possibly not be sample dependent because the processes are just due to the thermal activation, that could further be tested in the forthcoming investigations.

It should be noted that the 'temperature' is defined here as the 15 min duration of heating in the laboratory. The actual lithic heat treatment would probably be longer and cooling would probably take longer as the material size is larger than the one in the laboratory which is powder. The effect of heating would therefore be more, implying that the actual temperature should be less than the estimation made in the above protocol. The heating effect is the product of temperature and duration which can be quantitatively discussed as thermal activation processes. The duration of heat treatment should be further discussed in archaeological context for better estimation of the temperature.

7.4 The cases that the samples are old

7.4.1 The dose response of the E_1' center in quartz

The above discussions are valid when the heat treatment was made recently, as mentioned at the beginning of previous section. The word, 'recently', means the duration in which the natural radiation does not change the signal intensity of the E_1' center much enough to be detected. Toyoda and Amimoto (2020) investigated how gamma ray irradiation changes the signal intensity of the E_1' center. Their results are interesting that there are both cases, radiation reduces the intensity when the quartz has been preheated below 300 °C while, when preheated at 400 °C, increases at lower dose and decrease at higher dose (figure 7.8).

They attributed the phenomena to the electronic processes possibly occurring in quartz that radiation supplies electrons and heating supplies electronic holes to the oxygen vacancies and to the E_1' centers. According to their interpretation, as shown in figure 7.9, electronic holes are supplied to diamagnetic oxygen vacancies (with two electrons) on heating so that a diamagnetic oxygen vacancy turns a paramagnetic E_1' center after recombination of an electronic hole and one of the two electrons below 300 °C while an electronic hole is further recombine with the remaining electron to become an oxygen vacancy with no electron which shows no ESR signal. With irradiation, electrons are supplied so that an E_1' center turns a diamagnetic

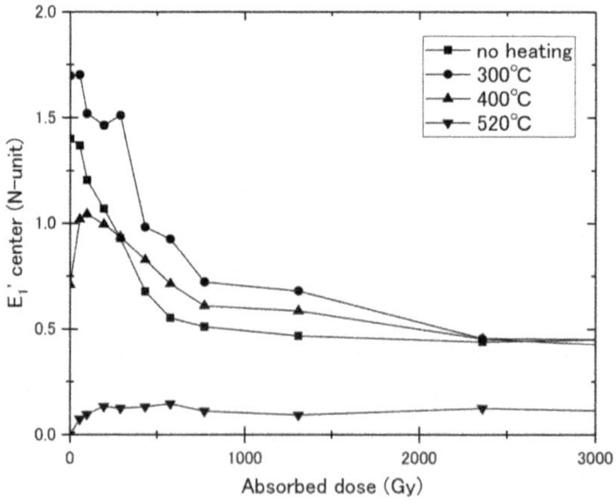

Figure 7.8. Dose responses of the E_1' center. Reproduced from Toyoda and Amimoto (2020) with permission from De Gruyter. The temperatures indicate the ones at which the samples were preheated before irradiation.

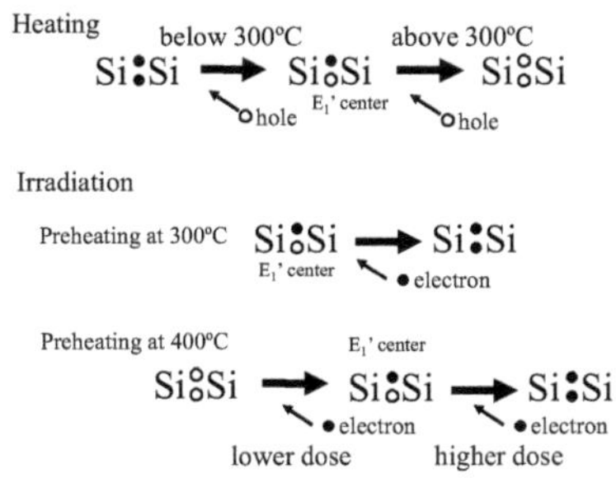

Figure 7.9. Electronic processes estimated to occur in quartz. Reprinted from Toyoda and Amimoto (2020) with permission from De Gruyter.

oxygen vacancy in the sample preheated at 300 °C. In the sample preheated at 400 ° C, an electron is supplied to an oxygen vacancy with no electrons to form an E_1' center at lower doses which increases the ESR intensity while an electron is further supplied which makes a formed E_1' center turn a diamagnetic oxygen vacancy with two electrons to result in decrease of the intensity at higher doses.

As a whole, heating and irradiation electronically induce the reactions in quartz to the opposite directions. Therefore, in older samples, natural radiation would cancel the effect at least partly of the heat treatment to the ESR signal. It would mean that the heating effect will be estimated to a degree lower than actual without considering the effect of radiation. Quantitative discussions on this issue would require detailed experimental investigations; however, qualitative rough discussions would be possible as in the following.

7.4.2 Estimation of the heat treatment of old lithic tools

As shown in figures 7.4 and 7.8, gamma ray doses in the range of 100 Gy will give changes in the signal intensity of the E_1' center. Therefore, the doses, 10 % of this amount, i.e. 10 Gy, would not give so much effect although it depends on how precisely the heat treatment temperature needs to be estimated. Assuming a natural dose rate of 3 mGy yr^{-1}, this dose corresponds to about 3000 years.

Samples with ages around 30 ka (cases B)

When the natural dose is about 100 Gy (samples with ages around 30 ka), the temperature estimation following the chart in figure 7.6 should not be correct because the natural radiation has already modified the original intensity of the E_1' center to a degree which cannot be neglected. However, it is possible to estimate what happens to these samples based on the laboratory experimental results already obtained. The chart in figure 7.10 will tell the temperature ranges in such samples.

If the temperature was below 300 °C (case B-0), even though the natural radiation has modified the intensity, the intensity of the E_1' center will increase on heating at 300 °C for 15 min to a level which corresponds to the number of oxygen vacancies (figure 7.3).

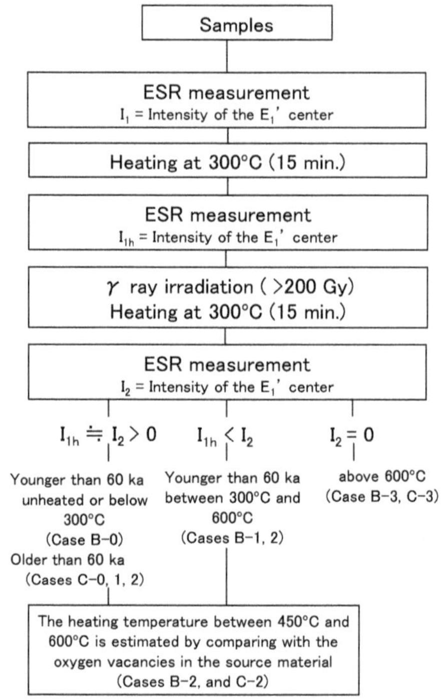

Figure 7.10. The protocol to estimate the temperature of lithic heat treatment proposed in this paper for old heated samples. See table 7.1 for heating categories.

On the other hand, if the temperature was between 300 °C and 600 °C (cases B-1, and 2), the intensity will increase on laboratory first heating (intensity I_{1h}) to a level lower than the one corresponding to the number of oxygen vacancies (figure 7.4). This will be examined by comparing this level with the one after gamma ray irradiation to 1 kGy and heating at 300 °C (intensity I_2). If the source material is available, the heating temperature may be quantitatively obtained by comparing the number of oxygen vacancies in the specimen with the one in the source material using the decay of the oxygen vacancies (figure 7.5) between 450 °C and 600 °C (case B-2).

If the temperature was above 600 °C (case B-3), no E_1' center will be observed in original samples nor in those after irradiation and heating because heating above 600 °C has annealed out the oxygen vacancies and the radiation doses are not enough to create oxygen vacancies.

Samples with ages around 60 ka and older (cases C)

When the natural dose is more than 200 Gy (samples with ages around 60 ka or older), the natural radiation has already modified the original intensity of the E_1' center significantly. In all cases that the temperature of the heat treatment was below 600 °C (cases C-0, 1, and 2), the intensity of the E_1' center will correspond to the number of oxygen vacancies after the first heating at 300 °C for 15 min (intensity I_{1h}).

This can be confirmed by observing the intensity of the E_1' center after gamma ray irradiation and following heating at 300 °C (intensity I_2). The heating temperature range cannot be estimated unless the number of oxygen vacancies is compared with the one in the source material in the temperature range between 450 °C and 600 °C (case C-2).

If the temperature was above 600 °C (case C-3), no E_1' center will be observed in original samples nor in those after irradiation and heating because heating above 600 °C has annealed out the oxygen vacancies and the radiation doses are not enough to create oxygen vacancies.

The protocol to estimate the temperature range

The protocol to estimate the temperature range of old lithic tools would be the following (figure 7.10). The first ESR measurement is not really meaningful but it is probably better to measure the intensity of the natural sample. The sample should be heated at 300 °C for 15 min before measuring the signal intensity, $I_{1\ \mathrm{h}}$, of the E_1' center. Then, after irradiation to more than 200 Gy and following heating at 300 °C for 15 min, the signal intensity, I_2, is obtained.

The signal intensities, $I_{1\ \mathrm{h}}$ and I_2, will be about the same when the sample was heated below 300 °C with an age younger than 60 ka (the natural dose less than 200 Gy) (case B-0) or below 600 °C with an age older than 60 ka (case C-0, 1). As for the sample older than 60 ka, one can just tell that the temperature was below (case C-0, 1 and 2) or above (case C-3) 600 °C, but not more precisely (cases C-0 to 2 cannot be distinguished). When $I_{1\ \mathrm{h}}$ is less than I_2, it means that the sample was heated younger than 60 ka (typically 30 ka) at a temperature between 300 °C and 600 °C. If a reference source material is available, the temperature may be estimated quantitatively by comparing the number of oxygen vacancies using figure 7.5 between 450 °C and 600 °C (cases B-2 and C-2). When the sample was heated above 600 °C (cases B-3 and C-3), I_2 will be zero as all oxygen vacancies have been annealed regardless of the age.

7.5 Sequence to be introduced into dating protocols

The method to estimate the heat treatment of lithic tools with ESR was proposed in 1993 (Toyoda *et al* 1993). Although it is not commonly used, this method would be as useful as other methods such as thermoluminescence in investigating the ancient technology of heat treatment as was used by Dunnell *et al* (1994). In addition, the author would like to emphasize that this method can be used in the procedures of dating of heat-treated lithic tools and pottery. Although there is only one work by Bartoll and Ikeya (1997) that impurity center (Al center) was used in dating of archaeological materials, it is necessary to show that heating has been completely erased the dating signal where the impurity centers would be most appropriate to be used for dating of quartz. The present method using the E_1' center and the oxygen vacancies would be useful for this purpose as was tried by Sayin *et al* (2020).

7.6 Summary

A method with electron spin resonance is presented in this chapter to assess the past heat treatment of lithic tools. It is to observe the E_1' center, a paramagnetic lattice defect in quartz, the method of which is based on physics, therefore quantitative discussions are possible.

When the heat treatment was made 'recently', the assessment is simple and would be reliable. The temperature ranges are easily categorized and the temperature can be determined if reference source material is available. However, for old samples, say more than 30 ka, the method will have much ambiguity because natural radiation would have modified the signal intensity, where further detailed investigations are needed to establish the method. The method with ESR would be useful in discussing the ancient technology for lithic tools and also in dating to check if the temperature of the heat treatment has been high enough to erase the dating signals.

References

Bartoll J and Ikeya M 1997 ESR dating of pottery: a trial *Appl. Radiat. Isot.* **48** 981–4

Collins M B and Fenwick J M 1974 Heat treatment of chert: Methods of interpretation and theirapplication *Plains Anthropol.* **19** 134–45

Delagnes A, Schmidt P, Douze K, Wurz S, Bellot-Gurlet L, Conard N J, Nickel K G, van Niekerk K L and Henshilwood C S 2016 Early evidence for the extensive heat treatment of silcrete in the Howiesons Poort at Klipdrift Shelter (layer PBD, 65 ka), South Africa *PLoS One* **11** e0163874

Dunnell R C, Ikeya M, McCutcheon P T and Toyoda S 1994 Heat treatment of Mill Creek and Dover cherts on the Malden Plain, Southwest Missouri *J. Archaeol. Sci.* **21** 79–89

Feigl F J, Fowler W B and Yip K L 1974 Oxygen vacancy model for the E_1' centre in SiO_2 *Solid State Commun.* **14** 225–9

Jani M G, Bossoli R B and Halliburton L E 1983 Further characterization of the E_1' center in crystalline SiO_2 *Phys. Rev.* B **27** 2285–93

Melcher C L and Zimmerman D W 1977 Thermoluminescence determination of prehistoric heattreatment chert artifacts *Science* **197** 1359–62

Nuttal R H D and Weil J A 1981 The magnetic properties of the oxygen-hole aluminum centers in crystalline SiO_2 I $[AlO_4]^\circ$ *Can. J. Phys.* **59** 1709–18

Pawlish L A and Sheppard P J 1983 Thermoluminescent determination of paleoindian heattreatment in Ontario, Canada *Am. Antiq.* **48** 793–9

Rudra J K and Fowler W B 1987 Oxygen vacancy and the E_1' center in crystalline SiO_2 *Phys. Rev.* B **35** 8223–30

Sayin U, Toyoda S, Isikd I, Bakkala G, Yucee U R, Kilica H S and Ozmena A 2020 ESR/OSL dating and firing temperature determination of archeological pottery from Kumyer Location in Turkey *Appl. Magn. Reson.* **53** 1361–74

Toyoda S and Amimoto M 2020 Dose response of the E_1' center in quartz *Geochronometria* **48** 191–6

Toyoda S and Hattori M 2000 Formation and decay of the E_1' center and of its precursor *Appl. Radiat. Isot.* **52** 1351–6

Toyoda S and Ikeya M 1991 Thermal stabilities of paramagnetic defect and impurity centers in quartz: basis for ESR dating of thermal history *Geochem. J.* **25** 437–45

Toyoda S, Ikeya M, Dunnell R C and McCutcheon P T 1993 The use of electron spin resonance (ESR) for the determination of prehistoric lithic heat treatment *Appl. Radiat. Isot.* **44** 227–31

Toyoda S, Rink W J, Schwarcz H P and Ikeya M 1996 Formation of E_1' precursors in quartz: applications to dosimetry and dating *Appl. Radiat. Isot.* **47** 1393–8

Toyoda S, Takeuchi D, Asai T, Komuro K and Horikawa Y 2005 Spin-spin relaxation times of the E_1' center in quartz with and without irradiation: implications for the formation process of the oxygen vacancies in nature *Radiat. Meas.* **39** 503–8

Weeks R A and Nelson C M 1960 Trapped electrons in irradiated quartz and silica: II. Electron spin resonance *J. Am. Ceram. Soc.* **43** 399–404

Weil J A 1984 A review of electron spin spectroscopy and its applications to thestudy of paramagnetic defects in crystalline quartz *Phys. Chem. Miner.* **10** 149–65

Chapter 8

Concordant ESR and TL depositional age of Romanian Plane loess

Liviu Tugulan, Mihail Secu, Vasile Bercu, Mihalis Cotrubinis and Octavian G Duliu

To determine the age of a 2.8 m loess horizon situated in the Eastern Romanian Plane, the independent age methods based on the accumulation of radiation defects, i.e. thermoluminescence (TL) and electron spin resonance (ES), were used for the same quartz fraction. In both cases, the additive method was used to determine the local paleodose (PD) while the local annual dose rate was recalculated starting from the content of radioactive elements in soil determined by high resolution gamma-ray spectrometry. Final results gave coincident results within experimental uncertainties of 19.85 ± 2.60 and 20.38 ± 2.93 ka for TL and respectively for ESR age. One possible explanation of the relatively high level of age uncertainties could be related to the additive method of the estimation of the PD by extrapolating the signal to dose dependency.

8.1 Introduction

One of the main problem in geochronology consists of dating recent (up to several Ma) sediments. Sedimentary material which can be deposited by the action of wind, water or ice, originates from much older rocks which, under the action of natural factors are disaggregated and transported at great distances. For this reason, sedimentary material itself has the age of its parent rocks, and not the age of its deposition, which makes the age methods based on the disintegration of natural radioactive elements existing in sediments useless.

This disadvantage could be overcome by using the accumulation of radiation centers produced by the natural radioactive elements existing in sediments. Indeed, the absolute majority of sedimentary material has the same average amount of natural radionuclides, closed to the upper continental crust (UCC), i.e. ^{40}K, ^{232}Th,

doi:10.1088/978-0-7503-4846-1ch8

^{235}U and ^{238}U (Zhang and Brady 2002, Rudnick and Gao 2003). The alpha, beta and gamma ray emitted by these elements generate at microscopic level dose rates between 5 and 20 mGy/a (Taylor and Aitken 1997, Tugulan and Duliu 2014). Moreover, the mean life time of these radionuclides varies between 708.3 Ma (^{235}U) and 14.05 Ga (^{232}Th), their presence maintain for long period of time a constant level of internal dose rate.

This generates in the composing minerals a multitude of structural defects, mainly consisting of trapped electrons by the impurity levels situated between valence and conduction bands. Their life time depends on the depth of trapping levels, so closer to valence band, longer their life time is, which, in some case, well overpasses 1 Ma (Ikeya 1993, Taylor and Aitken 1997).

This makes the trapped electrons excellent proxies for the sediments age, because at temperatures above 500 °C or exposed to solar light, they recombine with valence band holes, resetting the age clock.

The content of radiation trapped electrons can be determined with enough accuracy by thermoluminescence (TL) (Aitken 1985), optical stimulated luminescence (OSL) (Aitken 1998) and electron spin resonance (ESR) (Grün 1989, Ikeya 1993, Rink 1997).

Although intensively used in archaeology and quaternary geology, these methods have some drawbacks related to the long time stability of trapped electrons. Moreover, the circulation of highly soluble uranium carbonated could determine an uncontrolled long time variation of the annual dose rate (Ikeya 1993, Grün 1989).

This peculiarity was more evident in the case of ESR dating, whose accuracy is of the order of 10% and higher. For this reason, it is recommended any time is possible, the ESR dating to be check by a different geochronology method.

Accordingly, we have tried to determine the age of a loess deposit from Gorgana I archaeologic site located in the Romanian Plane, near the Radovana village, Calarasi County by TL and ESR and to check at which extent the ages provided by these methods are coincident within experimental uncertainties.

The results of this study will be further presented and discussed.

8.2 Materials and methods

8.2.1 Sampling and sample preparation

Loess and loess-like deposits in Romania are covering mainly the southern and southeastern part of the country along the lower Danube river (Conea 1969, Jipa 2013). The deposits can be up to ~50 m thick and contain several well-developed palaeosols thought to represent continuous and extended archives of regional climatic and environmental change during the Late and Middle Pleistocene (starting from 12 ka up to 800 ka).

The Gorgana I archaeologic site, Radovanu village (44.195°N, 26.513°E, Ialomita county, figure 8.1, inset), opened in 1920, is still active, a horizon laying at 3 m below the soil surface being investigated at present (Balescu and Morintz 2018). It contains the remains of an ancient Geto-Dacian fortress with earthen walls, some of them being recycled during Roman occupation (second century CE).

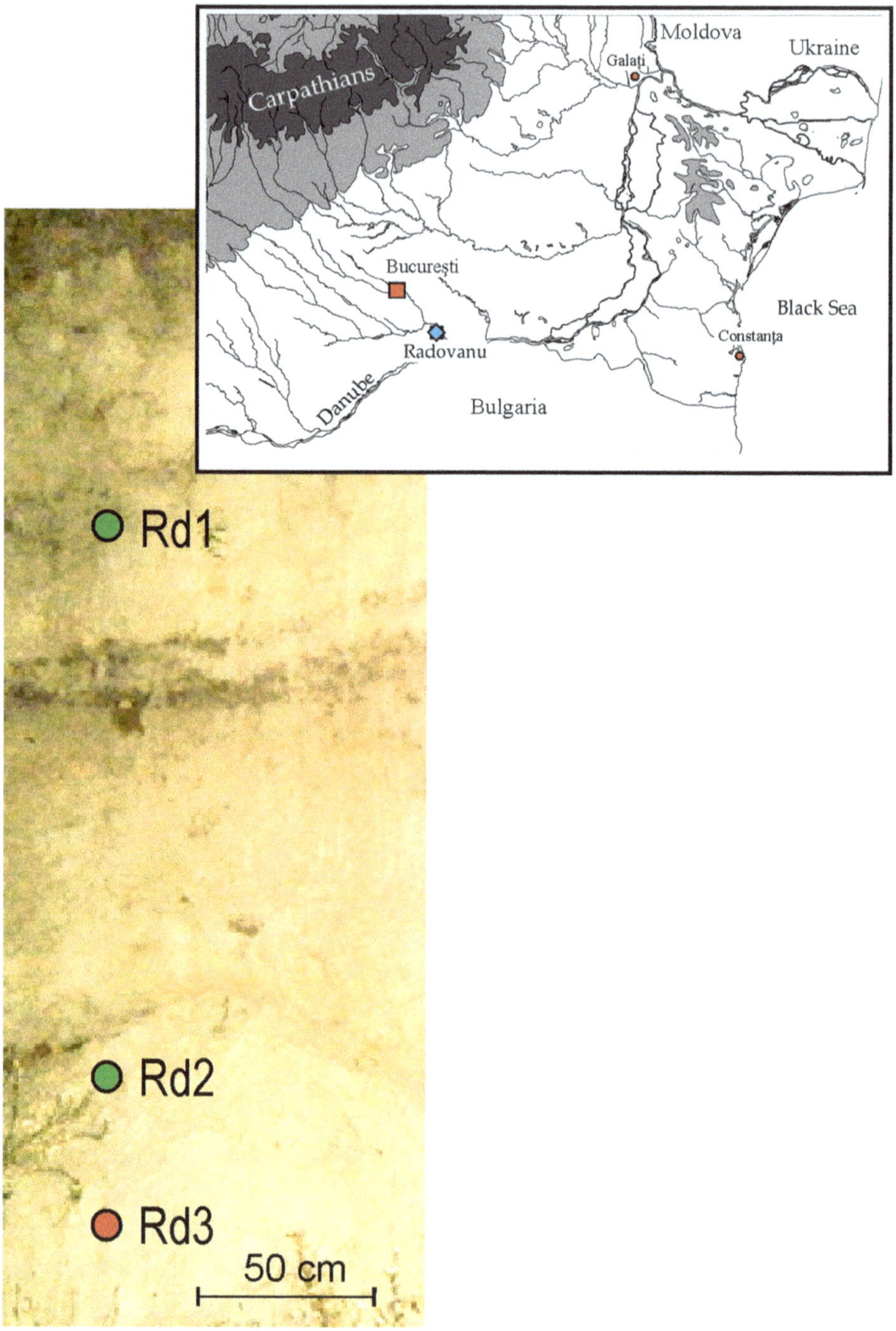

Figure 8.1. A section through Radovanu cliff showing the position of sampling points. The inset illustrates the geographical location of the Radovanu archaeological site.

The yellow brownish soil (figure 8.1), rich in loess, is characteristic for the Eastern Romanian Plain (Conea 1969, Jipa 2013).

For this study, at 2.8 m below surface, we have collected, by means of a 20 mm diameter, 30 cm long metallic corer, a soil sample consisting of about 2 kg of raw material. In laboratory, the soil was dried at constant mass and divided in three aliquots for further radiometric, TL, and ESR measurements.

Dried material of about 180 g were introduced in hermetically sealed plastic vials and kept four weeks to reach the radioactive equilibrium.

In the case of TL and ESR determinations, the quartz fraction showed to be the most adequate (Rink 1997). Accordingly, the silicates were separated from the carbonate and organic fractions by dissolving the soil sample in a solution of 1 N hydrochloric acid 10%, oxygenated water 10% and 80% distilled water. After complete dissolution of carbonates, the solution was filtered, the solid fraction being intensively washed with deionized water, dried at room temperature at constant weight and passed through a 170 and 270 mesh sieves to select the 63–90 μm grain fraction (Frechen *et al* 1996, Lang *et al* 1996).

By using a solution of sodium polytungstate, it was possible to select a mixture of quartz and plagioclase, the last one being removed by immersion in a solution of 40% hydrofluoric acid. This last treatment removed also a superficial layer of quartz grains containing the free radicals generated by the external alpha-rays (Lang *et al* 1996).

To avoid any errors induced by bleaching, beginning with the extraction of soil from corers and finishing with TL and ESR measurements, all operations were performed under protective red light.

8.2.2 High resolution γ-ray radiometric measurements

The content of natural radionuclides was determined, as mentioned before, by high resolution gamma-ray spectrometry. Accordingly, all gamma spectra were recorded for 36 000 s using a CANNBERA REGe detector, 40% efficiency, ISOCS™ calibrated (ISOCS 2020), provided with a 0.6 mm carbon-epoxy window and a DSA 1000 multichannel analyzer whose energy resolution was of 0.89 keV for the ^{57}Co 122 keV line and of 1.96 keV for the ^{60}Co 1332 keV line. The background was reduced by placing the detector inside of a 747 CANNBERA lead shield. Resulting gamma spectra were processed by using the GENIE™ 2000 basic spectroscopic software (GENIE 2000) while the detector efficiency calibration, self-absorption correction and coincidence-summing correction was done with LABSOCS™ (LABSOCS 2020) software. The background was estimated by measuring the spectrometer chain spectrum in the absence of any sample for $3 \times 3\,600\,000$ s.

At the same time, we have checked the existence of radioactive equilibrium. In the case of ^{232}Th this was done by comparing the intensity of ^{228}Ac gamma-ray line of with those of ^{212}Bi and ^{212}Pb while in the case of ^{238}U the radioactive equilibrium was checked by comparing the gamma spectrum of ^{234m}Pa with those of ^{214}Bi, ^{214}Pb, and ^{210}Pb (BIMP 2020). For both radioactive series, the considered activities were coincident within experimental uncertainties (Tugulan and Duliu 2014).

The accuracy of radiometric determinations was checked by means of the IAEA-RGTh-1 - Thorium ore (IAEA 2020) calibration reference material. To reduce the background due to radon accumulation, all measurements were done in working days under continuous ventilation (Tugulan and Duliu 2014).

Starting from the content of natural radionuclide determined as mentioned before, the annual dose rates were calculated by using the content-to dose rate coefficient provided by Guérin *et al* (2011). Under these conditions, the combined standard uncertainty (CSU) (Farrance and Frenkel 2012) was no greater than 6.1%.

It is worth mentioning that in order to check the homogeneity of loess, we have measured the dose rate for samples collected at three dept: 0.8 m (Rd1), 2.4 (Rd2) and 2.8 m (Rd3) (figure 8.1). For all sampling points the dose rates are coincident within one CSU, i.e. 3.32 ± 0.20, 3.41 ± 0.20 and 3.43 ± 0.21 mGy/a respectively.

8.2.3 ^{60}Co gamma ray irradiation

To determine the paleodose (PD) we have used the additive method which implied the irradiation of the quartz fraction with ^{60}Co gamma rays with increasing doses between 0 and 60 Gy (Ikeya 1993).

All irradiations were performed at room temperature using a Gamma Chamber 5000 (BRIT 2020) facility hosted by the Horia Hulubei National Institute for Physics Nuclear Engineering at a constant dose rate of 6210 Gy h^{-1}. An alanine dosimetry system was used to monitor the absorbed doses with a total uncertainty of ±5%. To attain the electronic equilibrium during irradiation, all samples were placed in 1.5 mm perspex tubes.

8.2.4 TL measurements

Thermoluminescence (TL) measurements were performed by using a Harshaw 3500 TL reader, in the 50 °C–410 °C temperature range with a heating rate of 4 °C s^{-1}. A monolayer of quartz grains (19 mg weight) was spread out on a 5 mm-diameter stainless steel disks and the TL radiation was recorded by using a blue transmitting BG12 filter. The resulting TL signal has a characteristic glow curve with a maximum centered above 310 °C, whose intensity was calculated as the area under the curve (figure 8.2(a)).

8.2.5 ESR measurements

The ESR measurements were carry out at room temperature using an X band ADANI CMS 8400 spectrometer provided with a TE_{102} rectangular resonant cavity and a 100 kHz modulation field with and 0.08 mT amplitude. Polycrystalline quartz fraction was introduced in 4 mm quartz tube, kept permanently in the same position within the resonant cavity. To increase the accuracy of the experimental analysis we have used a powder $CaO:Mn^{2+}$ reference sample permanently kept in the resonant cavity during measurements. For each sample, the magnetic field sweep field was extended to observe the third and the fourth line of the Mn^{2+} hyperfine structure ESR, used to normalize the amplitude of ESR quartz spectra (figure 8.2(b)). Each

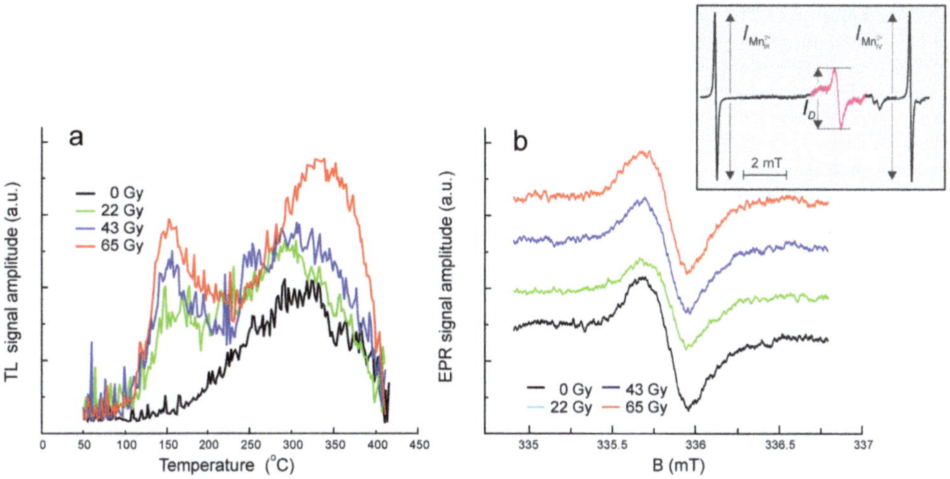

Figure 8.2. Dose dependencies of the quartz fraction glow curve (a) as well as of the ESR signal (b). The ESR line of the quartz fraction free radical is located between the third and the fourth Mn^{2+} hyperfine structure lines used as internal reference (b, inset).

sample was weighed before and after the ESR signal was recorded so that all ESR quartz spectrum lines were also normalized to sample mass.

$$I_{\mathrm{ESR}} = \frac{I_D}{0.5\left(I_{\mathrm{Mn}^{2+}_{III}} + I_{\mathrm{Mn}^{2+}_{IV}}\right)} \frac{1}{m} \tag{8.1}$$

where I_{ESR} represents the amplitude of the double normalized ESR sample line; I_D represents the amplitude of ESR line; m represents the sample mass; $0.5\left(I_{\mathrm{Mn}^{2+}_{III}} + I_{\mathrm{Mn}^{2+}_{IV}}\right)$ represents the average amplitude of the Mn^{2+} III and IV line reference signal (figure 8.2(b)).

In our case, the mass of the investigated samples (subscripts show the absorbed doses) were: $m_{0\ \mathrm{Gy}} = 98.7$ mg, $m_{20\ \mathrm{Gy}} = 41.9$ mg, $m_{40\ \mathrm{Gy}} = 43.9$ mg and $m_{60\ \mathrm{Gy}} = 54.6$ mg.

Periodically, the ESR signal of empty sample tubes was recorded to eliminate any parasitic signal (Bercu *et al* 2017).

As mentioned before, PD was determined using the additive method, i.e. after we have recorder the natural, unirradiated ESR signal, the sample was irradiated at different laboratory doses generating in this way the ESR response dose curve (Ikeya 1993).

8.3 Results and discussion

The local annual dose rate determined by taking into account, as mentioned before, the content of natural radionuclides was of 3.43 ± 0.21 mGy (table 8.1).

The PD was obtained from the linear dose dependence of the ESR or TL signals $I_{\mathrm{ESR/TL}}(D_{\mathrm{abs}})$ on the absorbed doses, according to the following expression:

Table 8.1. The experimental values of the annual dose rate, paleodose and corresponding TL and ESR age for the Radovanu 2.8 m loess horizon.

Method	Dose rate (mGy/a)	Paleodose (Gy)	Age (ka)
TL	3.43 ± 0.21	68.1 ± 8.0	19.85 ± 2.60
ESR		69.9 ± 9.2	20.38 ± 2.93

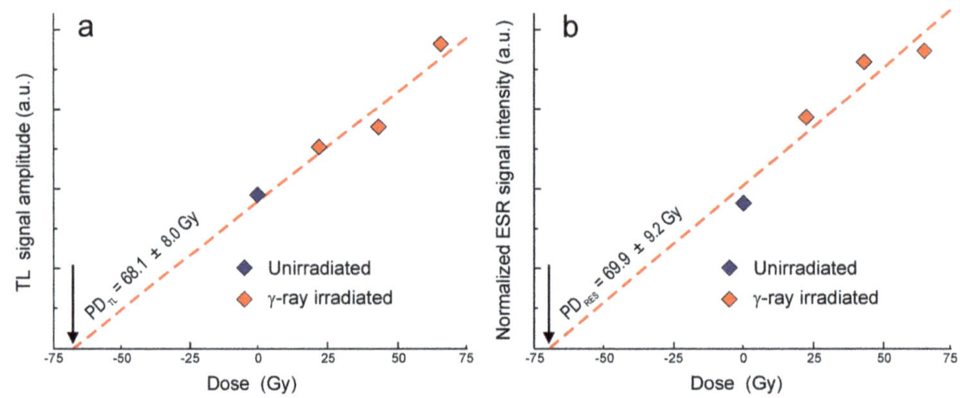

Figure 8.3. The determination of the paleodose (PD) by additive method in the case of TL (a) and ESR (b) geochronology. Black diamonds represent the initial signals while red diamonds represent the dose addition signal amplitudes.

$$I_{ESR/TL}(D_{abs}) = I_{ESR/TL}(0)\left[1 + \frac{D_{abs}}{PD}\right] \tag{8.2}$$

where $I_{ESR/TL}(0)$ represents the ESR/TL signal recorded for non-irradiated sample.

In these conditions the age T can be computed by using the annual dose rate Dr following the equation:

$$T = \frac{PD}{Dr}. \tag{8.3}$$

In the case of TL determinations, the signal as defined above, presented a linear dependence on absorbed dose (figure 8.3(a)) which gave a PD_{TL} of 68.1 ± 8 Gy. In these conditions, the soil age calculated by using the equation (3) for the annual dose rate of 3.43 ± 0.2 mGy yr^{-1} was of 19.85 ± 2.6 ka (table 8.1).

A closer result was obtain for ESR determinations. In this case, the ESR spectrum of quartz radiation centers reproduced in figure 8.2(b), consists of an almost symmetric line of which amplitude monotonously increases with the additive absorbed dose (figure 8.2(b)-inset, figure 8.3(b)). Following the same procedure, we have obtained the for PD_{ESR} = 69.90 ± 9.21 Gy from which, the soil age corresponding to the same annual dose rate was of 3.43 ± 0.2 mGy yr^{-1}, was of 20.38 ± 2.93 ka, in very good correlation with the TL age (table 8.1).

It should be remarked that in both cases, the ages, although coincident within one standard deviation, are affected by relatively high uncertainties of the order of 10%–15%, significantly higher than in the case of other absolute radiometric age methods based on the disintegration of long-lived natural radionuclides such as K–Ar, Ar–Ar, U–Pb, etc.

In our opinion, the source of a such higher uncertainty consists of the essence of additive dose technique where, the PD is determined by the intercept of extrapolated linear dependency of the TL or ESR signal amplitude produced by supplementary irradiation and not by an interpolation, as in the case of reconstructive techniques (Tsukamoto *et al* 2017). Despite this, our results showed a very good agreement between the two methods and a good concordance with Romanian Plane loess deposit age.

8.4 Conclusions

The additive reconstruction methods was used to determine by thermoluminescence (TL) and electron spin resonance (ESR) the age of a soil sample collected at depths between 0.9 and 2.8 m in an archaeological site located in the Eastern section of the Romanian Plain. The local values of dose rate, determined by gamma-ray spectrometry, were almost independent on the depth suggesting a homogeneous distribution of the main natural, long-life natural radionuclides.

The quartz fraction was selected progressively by means of acid dissolution and heavy liquid fractionation.

The absolute age was determined by TL as well as by ESR using the additive method applied to the quartz fraction, all supplementary irradiation being performed by ^{60}Co gamma rays.

TL was considered as the primary age method, while the ESR was used to check the validity of the TL results. Accordingly, the TL ages for the 2.8 m loess horizon was of 19.85 ± 2.6 ka, close to the ESR age of 20.38 ± 2.93 ka.

At the same time it is worth mentioning that both absolute age methods use the same kind of procedures according to which the paleodose is determined by the extrapolation of the signal vs dose dependency that could lead to relatively high age uncertainties.

Acknowledgements

This work was supported by a grant of the Romanian Ministry of Research and Innovation, CCCDI–UEFISCDI, project number PN-III-P1-1.2-PCCDI-2017-0686/52PCCDI/2018, within PNCDI III and partially accomplished within the cooperation protocol no. 4322-4-17/19 between JINR-Dubna, Russian Federation and the University of Bucharest.

References

Aitken M J 1985 *Thermoluminescence Dating* (London: Academic)

Aitken M J 1998 *An Introduction to Optical Dating: The Dating of Quaternary Sediments by the Use of Photon-stimulated Luminescence* (Oxford: Oxford University Press)

Balescu A and Morintz A 2018 A case of cynophagy at Radovanu—Gorgana a Doua (Calarasi County, 2nd–1st Centuryes BC) *Materials and Archaeological Researches, New series* **14** 133–48

Bercu V, Negut C D and Duliu O G 2017 Irradiation free radicals in freshwater crayfish *Astacus leptodactylus* Esch investigated by EPR spectroscopy *Radiat. Phys. Chem.* **133** 45–51

BIMP 2020 *Monographie BIPM-5 Table of Radionuclides. Bureau International des Poids et Mesures, Sèvres, France* https://bipm.org/en/publications/scientific-output/monographie-ri-5.html

BRIT 2020 *Gamma Chambers GC 5000* Board of Radiation & Isotope Technology, Department of Atomic Energy, Government of India http://britatom.gov.in/htmldocs/products06.html (Accessed 10 March 2020)

Conea A 1969 Profils de loess en Roumanie. La stratigraphy des loess d'Europe *Bulletin de l'Association Française pour l'étude du Quaternaire* ed J Fink (INQUA) pp 127–34

Farrance I and Frenkel R 2012 Uncertainty of measurement: a review of the rules for calculating uncertainty components through functional relationships *Clin. Biochem. Rev.* **33** 49–75

Frechen M, Schweitzer U and Zander A 1996 Improvements in sample preparation for fine grain technique *Ancient TL* **14** 15–7

GENIE 2000 *GENIE™ 2000 Basic Spectroscopic Software* https://mirion.com/products/genie-2000-basic-spectroscopy-software (Accessed 10 March 2020)

Grün R 1989 Electron spin resonance (ESR) dating *Quat. Int.* **1** 65–109

Guérin G, Mercier N and Adamiec G 2011 Dose-rate conversion factors: update *Ancient TL* **29** 5–8

IAEA 2020 IAEA-RGTh-1, Thorium ore reference material https://nucleus.iaea.org/sites/ReferenceMaterials/Pages/IAEA-RGTh-1.aspx (Accessed 10 March 2020)

Ikeya M 1993 *New Applications of Electron Spin Resonance* (Singapore: World Scientific)

ISOCS 2020 *ISOCS™ Calibration Software* https://mirion.com/products/s573-isocs-calibration-software (Accessed 10 March 2020)

Jipa D C 2013 Overview of the loess accumulation sedimentary features in the Romanian Plain and Dobrogea; S Radan, N Panin, D C Jipa and S-C Radan *Correlations of Quaternary Fluvial, Eolian, Deltaic and Marine Sequences: Field Trip Guidebook. INQUA e-Section on European Quaternary Stratigraphy (SEQS) Constanţa (Romania)*

LABSOCS 2020 *LABSOCS™ Callibration Software* https://mirion.com/products/labsocs-calibration-software (Accessed 10 March 2020)

Lang A, Lindauer S, Kuhn R and Wagner G A T 1996 Procedures used for optically and infrared stimulated luminescence dating of sediments in Heidelberg *Ancient TL l* **14** 7–11

Rink W J 1997 Electron spin resonance (ESR) dating and ESR applications in quaternary science and archaeometry *Radiat. Meas.* **27** 975–1025

Rudnick R L and Gao S 2003 Composition of the continental crust *Treatise on Geochemistry* vol 3 ed H D Holland and K K Turekian (Oxford: Elsevier-Pergamon) pp 1–64

Taylor R E and Aitken M J (ed) 1997 *Chronometric Dating in Archaeology* (New York: Springer)

Tsukamoto S, Porat N and Ankjærgaard C 2017 Dose recovery and residual dose of quartz ESR signals using modern sediments: implications for single aliquot ESR dating *Radiat. Meas.* **106** 472–6

Tugulan L C and Duliu O G 2014 Annual dose rate determination by high resolution gamma spectrometry for TL dating of loess deposits in south-eastern Dobrudjea, Romania *Rom. Rep. Phys.* **66** 862–74

Zhang P-C and Brady P V 2002 *Geochemistry of Soil, Radionuclides, Soil Science Society of America Special Publication Number* 59 (Madison, WI: Soil Science Society of America)

IOP Publishing

Spectroscopic and Microscopy Techniques for Archaeological and Cultural Heritage Research (Second Edition)

Ashutosh Kumar Shukla

Chapter 9

Electron spin resonance spectroscopy: a tool for dating mollusc shells, corals, and other materials

Puja Karmakar, Lokanath Mishra and Monalisa Mishra

Different dating techniques are used to determine the ages of samples. Among such techniques, electron spin resonance (ESR) dating has advantages for several reasons. ESR can determine the ages of hard materials, such as teeth. The ESR technique enables us to know the ages of materials, in particular the ages of fossils. Thus it helps to cast new light on human evolution. ESR can determine the ages of samples that are even more than 500 000 years old. In addition to teeth, the ages of molluscs can also be determined using the ESR technique. ESR can also be used to measure the ages of terrestrial dune sequences. The ages of corals can also be determined using the ESR technique. This chapter summarizes the history, advantages, and different sample preparation methods of ESR techniques.

9.1 Introduction

In certain materials, electrons and holes accumulate in the crystal lattice over time. When the mineral is exposed to radiation, these electrons are emitted into the environment. The radiated electrons can be used as as a means of dating in the ESR technique (Yoshida 1996). As a consequence, minerals act like natural dosimeters in the ESR process (Aydaş *et al* 2015). The age of a sample can be determined by calculating the dose released by a mineral in the past and the dose released by a mineral in recent past (Preusser *et al* 2003). Therefore, ESR is a very popular technique in archaeological study. In archeology, tooth enamel dating is often one of the most popular ESR applications. Other techniques, such as optically stimulated luminescence (OSL) and thermoluminescence (TL), are also used for age determination. However, ESR is more advantageous than these two techniques, as it can

doi:10.1088/978-0-7503-4846-1ch9

determine the ages of a variety of samples over a longer time range. The ESR technique is very useful for studying human evolution, as it can provide valuable information. The dating range of ESR is more than 500 000 years, which is better than that of the radiocarbon dating method (Blackwell 2006).

The first ESR study was performed by Ikeya in 1975 (Ikeya 1975). Many materials, such as fossilized tooth enamel, carbonate deposits (i.e. stalagmites), and quartz were subsequently investigated using ESR. ESR dating can determine the ages of samples ranging from a few hundred years to one million years (Ma) old. ESR can detect the ages of fossil carbonates which are even 100 Ma old. It is often used to detect the ages of quartz samples (Selepchi and Duliu 2007). It can detect the age of groups that cannot be detected using ^{14}C, uranium series, and luminescence methods. The ESR method makes use of zeroing progressions (such as pressure, heat, shock, and bleaching by sunlight). Different samples undergo significant changes as a result of zeroing progressions. Such changes are used as parameters to detect important geological and archaeological events of the quaternary period. Teeth are the hardest materials in the body (Mishra 2021) and are often found in paleontological sites. Stalagmites and quartz are hard minerals that are mainly found in geological sites (Aydaş *et al* 2015). Both these minerals commonly interact with teeth. As a consequence, both these minerals are used to detect the historical age of teeth.

The ESR technique sometimes fails to give accurate data due to fluctuations in the environment. Despite this disadvantage, the ESR technique is used to study environmental alterations such as landscape change, climate change, and temporal change. The ESR technique also provides useful information about human evolution. To obtain more consistent ages for these events, methodological improvements are continually being sought. In the study of the effects of long-term radiation, free radicals and natural carbonates play a significant role, which is particularly due to their occurrence in different varieties, starting with various natural limestones and marble and ending with invertebrate exoskeletons. For this reason, natural carbonates have been thoroughly investigated using the ESR method. In addition to these variations, these methods have also been used at different geographical locations (Selepchi and Duliu 2007). This chapter describes ESR dating in terms of its physical background, sample selection, treatment, and analysis.

9.2 A brief history of electron spin resonance dating

In 1936, the principles of ESR spectroscopy were first described by Gorter and his colleagues (Gorter and Kronig 1936). Although ESR is similar in application to thermoluminescence, early attempts to date different substances were unsuccessful. In 1975, Ikeya first dated a stalagmite from Akiyoshi Cave in Japan (Ikeya 1975). After that, many researchers performed geochronological studies of fossils using dried blood, quartz, and engine oil. Some previous important applications included the dating of fault gouge, teeth, burnt flint, and bones. The early uses of this technique were unacceptable due to its inaccurate application (Aydaş *et al* 2015). Today, 60 ESR laboratories are established worldwide, among which, 25 are based in Japan. Those laboratories are researching ESR dating and dosimetry. Only about ten laboratories are dedicated to performing dating as routine work. In the day-to-day application of ESR, samples such as tooth enamel, molluscs, corals, and quartz from fault gouges are investigated. In addition, food irradiation and retrospective dosimetry are used for numerous basic studies that may lead to new geological and paleontological uses. The combination of advanced microscopy and imaging techniques with ESR is benefiting the development of new mineralogical and paleontological applications (Preusser *et al* 2003).

9.2.1 Techniques used for dating

9.2.1.1 Common techniques used for dating

Many methods are used to measure the age of geological and archaeological samples (Yoshida 1996). These methods depend on the rate of change of radioactive isotopes. Some of the radioisotopes include $^{40}K/^{40}Ar$, $^{40}Ar/^{39}Ar$, and $^{230}Th/^{234}U$. The decay of cosmogenically created radionuclides (for example, ^{14}C) is used to monitor the ages of samples. It also detects the rate of accumulation of radiation damage in crystals, cause and using, for example, thermoluminescence, optically invigorated luminescence, ESR, and fission tracks. Global time markers such as paleomagnetism and stable oxygen isotopes are also taken into consideration. The rate of chemical changes of organic and inorganic materials, for example amino acid racemization, obsidian hydration, and cation ratios are also considered in this process (Aydaş *et al* 2015). Among these strategies, only a few can be used for the kinds of materials typically observed in Quaternary sites (Selepchi and Duliu 2007, Yoshida 1996).

9.2.1.2 The ^{14}C methods and their disadvantages

Coastal deposition mainly comprises quartz and shell hash. Since other materials are not accessible for dating, carbonate composites comprising shell pieces have been broadly dated using ^{14}C to determine the deposition time of marine silt (Chapman *et al* 2016, Thom *et al* 1981, Thom and Roy 1985). To obtain the carbon required for a ^{14}C examination, a minimum of 30 g of shell sample is required (Gupta and Polach 1985). Alternatively, a milligram-scale sample can be subjected to accelerator mass spectroscopy (AMS), which can directly date the shell sections (Selepchi and Duliu 2007). This is not viewed as a practical technique, however, due to the significant expense of AMS dating. Moreover, a restricted dating range is an issue for both ordinary and AMS ^{14}C strategies. The half-life of ^{14}C is 5730 ± 40 years, which in principle allows age to be determined beyond 40 000–50 000 years (Preusser *et al* 2003).

Recent improvements in AMS strategy have stretched this hypothetical limit to 70 000 years. In practical terms, however, the limit is often significantly lower (30 000–40 000 years) due to *in situ* contamination of old samples by modern carbon (Aitken 1991). Likewise, ^{14}C ages need to be aligned with historical changes in the production rate of ^{14}C in the upper atmosphere. Reliable calibration information is accessible for the last 11 000 years (Guilderson *et al* 2005), but the dependability is reduced for ages of up to 30 000 years (Bard *et al* 1998). The 30 000 years' working limit of ^{14}C dating is far too short to cover the series of geological and archaeological events that happened during the 2 Ma Quaternary period (Selepchi and Duliu 2007, Yoshida 1996).

9.2.2 Other absolute dating methods and their problems

9.2.2.1 Uranium-series methods

The most widely used uranium-series dating is the method that depends on the disequilibrium in the ^{238}U decay series, which includes the changing ratio of two radioactive daughter products, i.e. ^{234}U and ^{230}Th. This technique can be utilized to determine ages up to around 400 K annum (ka) (Ivanovich and Harmon 1992). It has been applied most effectively to fossil corals (Ashby *et al* 1987), which appear to have acted as chemically closed since the hour of their growth. On the other hand, the use of ^{238}U and its daughter products to date fossil molluscs and sea urchins has not been appropriate, due to their closed structures (Kaufman 1971, Muhs and Kennedy 1985, Szabo and Vedder 1971). In some places, the late Quaternary sediments mainly consolidate fossil molluscs, and these probably behaved as an open system. At this stage, it is not possible to determine the history of the U-gain and loss of these mollusc shells. As a consequence, uranium-series dating cannot be applied to such deposits today. However, alpha spectrometric and mass spectrometric U-series ages have been acquired from the raised coral reef porches on the Huon Peninsula, Papua New Guinea (Selepchi and Duliu 2007).

9.2.2.2 Luminescence methods

TL and OSL techniques have been broadly used to date archaeological and geological incidents in the Quaternary, utilizing minerals such as quartz, feldspar, and zircon. The upper age limit for quartz has been shown to be more than 800 ka (Huntley *et al* 1993). The principle of the luminescence techniques depends on the time-dependent accumulation of trapped electrons and holes caused by normal radiation. Radiative recombination (related to the discharge of photons) is instigated by heating (TL) or illumination (OSL). The two methods allow the calculation of the time passed since the mineral grains were last exposed to sunlight (Wintle and Huntley 1979).

Minerals exposed to natural radiation accrue trapped unpaired electrons plus positively charged 'holes.' This characteristic of minerals produces signals that are detectable by ESR spectrometers. These signals are the results of flaws in the crystalline structure, which are related to trace contaminants (Aydaş *et al* 2015). Such signals have been reported using an energy band model for some minerals, such as quartz, apatite, and calcite. In insulators, there are two energy levels at which electrons can be present: the ground state, known as the valence band, and a higher-energy state known as the conduction band. In between these two energy bands, there is a forbidden energy zone (Selepchi and Duliu 2007). Radioactive isotopes that are naturally present emit different types of emission, which generally ionize atoms. Electrons, which are negatively charged, are removed from atoms in the ground state (the valance band), while positively charged holes persist in the valence band. After a short period, most of the electrons reunify with holes, and the mineral returns to its original form, i.e. an electrically neutral form. Natural minerals have flaw sites, for example lattice defects and interstitial atoms, which are places that can trap electrons. These places act as paramagnetic centers which can be identified using their characteristic ESR signals by an ESR spectrometer (Radtke *et al* 1985). During sample preparation, the mineral has no trapped electrons. As a consequence, the ESR signal intensity remains at zero. Later, during preparation, the mineral is subjected to natural radiation (α-particles, β-particles, and γ-rays). During this step, the mineral forms trapped electrons and holes. This step is continued until the sample is tested in the laboratory. The ESR signal obtained is known as the natural intensity; it is reliant on the quantity of traps, the strength of the radioactivity (dose rate, D), and the time of radiation T (one quarter of the sample's age). The intensity of the ESR signal is proportional to the number of centers produced. The inverse of the dose rate and the time is the dose, D_E, to which the sample was exposed at an earlier time (Schellmann and Radtke 1999). The age of the compound can be calculated from the formula below:

$$T = \frac{D_E}{D}.$$

D_E is the real ESR part of the dating process. The dose rate is (Radtke *et al* 1985) determined by an investigation of the radioactive elements such as Th, U, and K present within the sample as well as its surroundings. The amounts of radioactive elements are changed to dose rates using published tables. The strength of the normal radiation field that a sample exhibits is carefully evaluated in relation to the complexity of its structure.

9.2.3 The preparation of mollusc shell samples for electron spin resonance spectroscopy

The coastline of India has rich mollusc diversity (Pawar 2012). A prototypical ESR sample needs to be perfect and primeval. The removal of soil is a necessary step in this procedure, so the samples are washed with water to remove any soil. After cleaning, the samples are soaked in acetic acid for some minutes to an hour to eliminate the involvement of any external α-rays from the sediment. They are then again washed in distilled water and desiccated at 25 °C or 77 °F. Afterwards, the samples are gently crushed in a mortar, and the stress grains in the range of 125–250 μm are isolated. The isolated grains are further soaked in acetic acid for a little time to eliminate surface flaw signals at $g = 2.0002$ produced through compression during grinding (Engin *et al* 2006). Finally, the isolated grains are cleaned in distilled water for some minutes and desiccated at 25 °C. The grains are separated into groups, each of which weighs 150 mg. The entire sample preparation procedure is executed under dim red light. At the time of sample preparation, the criteria shown below should not be taken into account.

(I) Sticky substances can enhance any U impurity in the sample, thus decreasing the precision of inner dose rate measurements. In addition, organic composites can cause ESR signal interference.
(II) Cleaning and rinsing may eliminate U, datable mineral, and sediment. The sediment associated with the sample may offer the only chance to detect the peripheral dose rates.
(III) If used to eradicate samples from surfaced sediment, an acid suspension can soften the sample and filter its U.
(IV) If the attached sediments are removed, this decreases the accuracy of external dose measurements.
(V) If the samples are numbered using paints and inks, then there is a chance of contaminating the organic compounds.
(VI) Permitting mud samples to dry then requires wholesale crushing as part of the preparation process, which can somewhat bleach some ESR signals.
(VII) Packing samples for transport with substances such as old newspapers, tinted papers, etc. can cause some degree of organic contamination if the packing materials contact the samples. The best way to pack samples is to use inexpensive, undyed, unperfumed toilet paper (Blackwell 2006).

9.2.4 The preparation of coral samples

Coral reefs are decreasing at a frightening pace all over the world due to the cumulative diversity, regularity, and action of human influences (Hughes *et al* 2017, Lesser 2013). Consequently, the persistence and prevalence of worldwide combined studies of subsidiary coral reef organization, restoration, and aquaculture efforts have been enhanced (Hughes *et al* 2017). Scleractinian corals are frequently used as biomonitors for phenological and environmental applications in tropical reef ecosystems, given their extraordinary sensitivity to physical and biochemical deviations in aquatic environs (Carlos Filho *et al* 2012, Goffredo *et al* 2011).

In recent times, the method most commonly used to date coral samples is U–Th dating. However, if we go beyond the recent timeline and examine the past, ^{14}C dating can be used. Thus, the question arises of which is better and which gives the maximum data about the age of corals? A combination of both techniques is often used, since each technique has its own drawbacks when applied to coral samples. The marine reservoir effect is a case in which the long-standing carbon in the deep ocean can give a false radiocarbon age, leading to doubts. In contrast, U–Th dating can be less accurate, and if thorium is present as a secondary source in the coral structure, it becomes difficult to use U–Th dating.

Corals take up strontium isotopes and incorporate them into their structure; they keep such oceanic strontium throughout their whole lives. Their age can be determined by measuring the $^{87}Sr/^{86}Sr$ ratio in a sample. If an assumption is made about typical sea water, the age determination can be completed. In radiocarbon dating sample calibration is difficult due to the presence of strontium (El Meknassi *et al* 2018). Faced with the problem of coral dating, several dating methods can be combined together as a means of problem solving and eliminating uncertainties (Plank and Forsyth 2016). Samples can also be collected by dredging. Samples can be identified as modern or old using their intact organic matter. The core of the coral sample is placid; the samples should be strained to obtain an element size of 120–250 mm. The samples need to be tested by x-ray diffraction and ESR screening before dating takes place.

9.2.5 Sample cutting for electron spin resonance dating

Using a thin-walled theca that connects the septa, which are radially symmetric, samples were divided into small slabs. The septa of particular deep-sea corals were categorized based on their comparative cross-sectional areas in a circular plane. The slabs were chopped and then separated for one of two cleaning processes, which are detailed below.

9.2.5.1 Ultrasonic cleaning

Samples were examined using a binocular microscope for the presence of organic coatings or iron–manganese oxide crusts. These coatings were scraped off using dental tools. The samples were placed in a plastic bottle, cleaned with an ultrasonic cleaner for ten minutes in deionized water, and then thoroughly rinsed. This process was repeated multiple times until the coral appeared clean (crust-free) under the microscope. The aforementioned technique was then carried out thrice in a Teflon beaker, each time followed by a 5 min ultrasonic cleaning. The sample was then prepared for isotopic analysis and dried in an oven at around 70 °C.

9.2.5.2 Chemical cleaning

With the exception of changing the pre-cleaning step to more effectively remove the black crusts, the approach outlined by Shen and Boyle (1988) was followed (Shen and Boyle 1988). The samples were cleaned using a brush and deionized water, then put in polypropylene tubes with clean, distilled water and ultrasonically sonicated for 15 min. The corals were then immersed in a 50/50 solution of 30% H_2O_2 and 1 M NaOH and subjected to sonication for 15 min. This step was repeated numerous times until there was little crust remaining. In order to remove any loose material, samples were occasionally wiped with a brush in between oxidative leaches. These oxidizing procedures make it simple to remove the observable iron–magnesium oxide crusts, which flake off in minute sheets. This implies that an organic 'glue'—possibly organic debris from polyps—holds the crust together. Following Shen and Boyle's approach from 1988, subsequent cleaning was conducted using only clean reagents.

9.3 Thermal ionization mass spectrometry

Samples that had been chemically or ultrasonically cleaned were progressively dissolved in nitric acid. All samples were entirely dissolved by this, with the exception of the iron–manganese oxide crust samples, which still contained some residue. The leftover material, most likely detrital aluminosilicate present in the crust, rapidly solubilized in hydrofluoric acid. The solutions were subsequently spiked with known concentrations of ^{229}Th and ^{233}U–^{236}U ($^{233}U/^{236}U$ 51.010 527). After the thorium and uranium were chemically separated, the thorium fraction was put onto a graphite-coated single Re filament whose thorium blank had already been verified. At about 1700 °C, the filament blank for ^{232}Th typically ranges from 100 to 150 counts per second, and the count rate gets faster as the temperature rises. In typical sample runs, this is a sizeable portion of the overall ^{232}Th beam, especially for tiny samples with low ^{232}Th concentrations. The blank was adjusted by assuming a 675% variability in its value. ^{232}Th was measured at temperatures below 1700 °C to reduce the filament blank, but, as previously demonstrated (Leal *et al* 2016), the ambiguity in the filament blank correction is the primary cause of error in ^{232}Th results. The uranium fraction was loaded onto a Re filament devoid of graphite, and the double-filament method was used to run the experiment. At the first stage of electron multiplication before the stationary quadrupole second stage, the Finnigan-MAT 262-RPQ mass spectrometer at the Minnesota Isotope Laboratory detected uranium and thorium. A measurement of count rates at half masses was used to account for trials.

9.4 Information obtained from different electron spin resonance analyses

The assessment of ESR and U–Th age result approximations has confirmed that they display noteworthy reliability. This reliability depends on the primary acceptance of U in the tested corals, and henceforth the custom of the early-interest prototypical for ESR phase calculations. For ESR to become the prevalent practice for Quaternary study, it now seems necessary that the development of ESR should achieve the lowest possible level of adjustment for the two most predominant ESR dating requests (i.e. fossil tooth enamel and quartz). These methods are habitually requested by many workshops all over the world. Efforts have nearly been completed in this regard for ESR data broadcasting (Duval *et al* 2018, Grün 1992). Nonetheless, this cannot be accomplished without the sustained support of the ESR dating community and the noble evaluation of papers (Radtke *et al* 1985, Timar-Gabor *et al* 2020). A major issue related to the applicability of this approach to mollusc shell dating is whether this substance, like tooth enamel, can be regarded as an open system for U-series assessment. An ESR age measurement was performed by Schellmann *et al* 2020 (Schellmann *et al*, 2020). Their method mainly used the developmental absorption model (= assuming a closed system). The total agreement with impartial age control obtained by this study may be construed as a subsidiary indication that most mollusc shells behave as closed systems (Molodkov 1993, Schellmann *et al* 2020). The aragonitic shell of a mollusc is likely to be the most common of all ESR dating substances. Admittedly, due to their open-system behavior with respect to uranium, shell particles must be reviewed for transposition from aragonite to calcite, which is the very same issue that plagues U-series shell dating. Moreover, species-specific identification is necessary in order to implement ESR dating (Schellmann *et al* 2020, Yoshida 1996). Nonetheless, the first systematic dating analysis of shells was carried out by Radtke *et al* (1985). Ikeya and Ohmura also focused on the ESR dating of shells (Ikeya and Ohmura 1981, Preusser *et al* 2003, Yoshida 1996).

9.5 Conclusions

ESR dating is predominantly used for the dating of teeth. Tooth dating using the ESR technique, particularly for the purpose of dating human fossils, adds to our knowledge of the evolution of modern humans. ESR has a dating range which extends well beyond that of radiocarbon and covers the last 500 000 years. Mollusc $^{230}Th/^{234}U$ ages can be hard to acquire, particularly if deposition has impacted the samples. ESR can be a valuable tool for the dating of terrestrial dune sequences. Furthermore, no single dating method, either the use of $^{230}Th/^{234}U$ on corals or the use of ^{14}C on organic matter, can be considered infallible unless confirmed by an independent method. Consequently, ESR can provide critical information about the ages of terrestrial and marine fossil shells. The samples' dose–response characteristics are very likely to match those of other carbonate samples (Molodkov 1993).

In short, ESR is an indispensable tool in dating fossils.

References

Aitken M 1991 Principles of radioactive dating *Scientific Dating Methods* (Dordrecht: Kluwer Academic Publishers) pp 3–13

Ashby J R, Ku T and Minch J A 1987 Uranium series ages of corals from the upper Pleistocene Mulege terrace, Baja California Sur, Mexico *Geology* **15** 139–41

Aydaş C, Engin B, Kapan S, Komut T, Aydın T and Paksu U 2015 Dose estimation, kinetics and dating of fossil marine mollusc shells from northwestern part of Turkey *Appl. Radiat. Isot.* **105** 72–9

Bard E, Arnold M, Hamelin B, Tisnerat-Laborde N and Cabioch G 1998 Radiocarbon calibration by means of mass spectrometric 230Th/234U and 14C ages of corals: an updated database including samples from Barbados, Mururoa and Tahiti *Radiocarbon* **40** 1085–92

Blackwell B A 2006 Electron spin resonance (ESR) dating in karst environments *Acta Carsologica* **35** 123–53

Carlos Filho V, dos Santos W P, de Kikuchi R K and Hatje V 2012 Decontamination procedure for trace elements determination in coral skeleton samples *Microchem. J.* **102** 123–9

Chapman A D, Jacobson C E, Ernst W, Grove M, Dumitru T, Hourigan J and Ducea M N 2016 Assembling the world's type shallow subduction complex: detrital zircon geochronologic constraints on the origin of the Nacimiento block, central California Coast Ranges *Geosphere* **12** 533–57

Duval M, Grün R, Parés J M, Martín-Francés L, Campaña I, Rosell J, Shao Q, Arsuaga J L, Carbonell E and de Castro J M B 2018 The first direct ESR dating of a hominin tooth from Atapuerca Gran Dolina TD-6 (Spain) supports the antiquity of Homo antecessor *Quat. Geochronol.* **47** 120–37

El Meknassi S, Dera G, Cardone T, De Rafélis M, Brahmi C and Chavagnac V 2018 Sr isotope ratios of modern carbonate shells: good and bad news for chemostratigraphy *Geology* **46** 1003–6

Engin B, Kapan-Yeşilyurt S, Taner G, Demirtaş H and Eken M 2006 ESR dating of Soma (Manisa, West Anatolia–Turkey) fossil gastropoda shells *Nucl. Instrum. Methods Phys. Res., Sect.* B **243** 397–406

Goffredo S, Vergni P, Reggi M, Caroselli E, Sparla F, Levy O, Dubinsky Z and Falini G 2011 The skeletal organic matrix from Mediterranean coral *Balanophyllia europaea* influences calcium carbonate precipitation *PLoS One* **6** e22338

Gorter C and Kronig R L 1936 On the theory of absorption and dispersion in paramagnetic and dielectric media *Physica* **3** 1009–20

Grün R 1992 Remarks on ESR dating of fault movements *J. Geol. Soc.* **149** 261–4

Guilderson T P, Reimer P J and Brown T A 2005 The boon and bane of radiocarbon dating *Science* **307** 362–4

Gupta S K and Polach H A 1985 *Radiocarbon dating practices at ANU* (Canberra: Australian National University)

Hughes T, Kerry J, Álvarez-Noriega M, Álvarez-Romero J, Anderson K, Baird A, Babcock R, Beger M, Bellwood D and Berkelmans R 2017 Global warming and recurrent mass bleaching of corals *Nature* **543** 373–7

Huntley D, Hutton J and Prescott J 1993 Optical dating using inclusions within quartz grains *Geology* **21** 1087–90

Ikeya M 1975 Dating a stalactite by electron paramagnetic resonance *Nature* **255** 48–50

Ikeya M and Ohmura K 1981 Dating of fossil shells with electron spin resonance *J. Geol.* **89** 247–51

Ivanovich M and Harmon R S (ed) 1992 *Uranium-Series Disequilibrium: Applications to Earth, Marine, and Environmental Sciences* 2nd edn (Harwell: AEA Technology)

Kaufman A 1971 U-series dating of Dead Sea basin carbonates *Geochim. Cosmochim. Acta* **35** 1269–81

Leal M C, Ferrier-Pagès C, Petersen D and Osinga R 2016 Coral aquaculture: applying scientific knowledge to ex situ production *Rev. Aquac.* **8** 136–53

Lesser M 2013 Using energetic budgets to assess the effects of environmental stress on corals: are we measuring the right things? *Coral Reefs* **32** 25–33

Mishra M (ed) 2021 Self-assembly of teeth proteins and their applications in dentistry *Nanotechnology for Dentistry Applications* (Bristol: IOP Publishing) pp 12 12–1

Molodkov A 1993 ESR-dating of non-marine mollusc shells *Appl. Radiat. Isot.* **44** 145–48

Muhs D R and Kennedy G L 1985 An evaluation of uranium-series dating of fossil echinoids from southern California Pleistocene marine terraces *Mar. Geol.* **69** 187–93

Pawar P R 2012 Molluscan diversity in mangrove ecosystem of Uran (Raigad), Navi Mumbai, Maharashtra, west coast of India *Bull. Environ. Pharmacol. Life Sci.* **1** 55–9

Plank T and Forsyth D W 2016 Thermal structure and melting conditions in the mantle beneath the Basin and Range province from seismology and petrology *Geochem. Geophys. Geosyst.* **17** 1312–38

Preusser F, Radtke U, Fontugne M, Haghipour A, Hilgers A, Kasper H U, Nazari H and Pirazzoli P 2003 ESR dating of raised coral reefs from Kish Island, Persian Gulf *Quat. Sci. Rev.* **22** 1317–22

Radtke U, Mangini A and Grün R 1985 ESR dating of fossil marine shells *Nucl. Tracks* **10** 879–84

Schellmann G and Radtke U 1999 Problems encountered in the determination of dose and dose rate in ESR dating of mollusc shells *Quat. Sci. Rev.* **18** 1515–27

Schellmann G, Schielein P, Burow C and Radtke U 2020 Accuracy of ESR dating of small gastropods from loess and fluvial deposits in the Bavarian Alpine Foreland *Quat. Int.* **556** 198–215

Selepchi E and Duliu O 2007 Comparative study on ESR spectra of carbonates *Rom. J. Phys.* **52** 657–66

Shen G T and Boyle E A 1988 Determination of lead, cadmium and other trace metals in annually-banded corals *Chem. Geol.* **67** 47–62

Szabo B J and Vedder J G 1971 Uranium-series dating of some Pleistocene marine deposits in southern California *Earth Planet. Sci. Lett.* **11** 283–90

Thom B, Bowman G and Roy P 1981 Late quaternary evolution of coastal sand barriers, Port Stephens-Myall Lakes area, central New South Wales, Australia *Quat. Res.* **15** 345–64

Thom B G and Roy P S 1985 Relative sea levels and coastal sedimentation in southeast Australia in the Holocene *J. Sediment. Res.* **55** 257–64

Timar-Gabor A, Chruścińska A, Benzid K, Fitzsimmons K, Begy R and Bailey M 2020 Bleaching studies on Al-hole ($[AlO_4/h]^0$) electron spin resonance (ESR) signal in sedimentary quartz *Radiat. Meas.* **130** 106221

Wintle A and Huntley D 1979 Thermoluminescence dating of a deep-sea sediment core *Nature* **279** 710–12

Yoshida H 1996 Quaternary dating studies using ESR signals, with emphasis on shell, coral, tooth enamel and quartz *PhD Thesis* Australian National University

www.ingramcontent.com/pod-product-compliance
Ingram Content Group UK Ltd.
Pitfield, Milton Keynes, MK11 3LW, UK
UKHW051041120926
8556IPUK00030B/634

9 780750 348447